D1519765

# Range-Doppler Radar Imaging and Motion Compensation

DISCLAIMER OF WARRANTY



For a listing of recent titles in the *Artech House Radar Library*,
turn to the back of this book.

# Range-Doppler Radar Imaging and Motion Compensation

Jae Sok Son
Gabriel Thomas
Benjamin C. Flores

Artech House
Boston • London
www.artechhouse.com

**Library of Congress Cataloging-in-Publication Data**
Son, Jae Sok.
Range-Doppler radar imaging and motion compensation / Jae Sok Son, Gabriel Thomas, Benjamin C. Flores
p. cm. — (Artech House radar library)
Includes bibliographical references and index.
ISBN 1-58053-102-4 (alk. paper)
1. Doppler radar. 2. Image compression. I. Thomas, Gabriel. II. Flores, Benjamin C. III.Title. IV. Series.

TK6592.D6 S66 2000 00-064281
621.3848—dc21 CIP

**British Library Cataloguing in Publication Data**
Son, Jae Sok
Range-Doppler radar imaging and motion compensation. —
(Artech House radar library)
1. Doppler radar 2. Moving target indicator radar
I. Title II. Thomas, Gabriel III. Flores, Benjamin C.
621.3'848

ISBN 1-58053-102-4

**Cover design by Christina Stone**


**685 Canton Street**
**Norwood, MA 02062**



International Standard Book Number: 1-58053-102-4
Library of Congress Catalog Card Number: 00-064281

10 9 8 7 6 5 4 3 2 1

# Contents

# Preface

The last decade of the 20th century saw intense activity in the development, testing, and implementation of algorithms for range-Doppler imagery. Nowadays, a growing research community of government leaders, industry researchers, and academics attends and participates in international conferences dedicated to this field of study. In writing this book, we have condensed a significant body of work in the area of motion compensation as it relates to the field of inverse synthetic aperture radar (ISAR) mapping, which is a particular form of range-Doppler imaging. Our purpose is to introduce the practicing radar engineer to some of the advanced methods used to analyze and synthesize the echo transfer functions of ISAR targets.

The concepts and algorithms disclosed here were developed with the assistance of a talented group of graduate students at the University of Texas at El Paso, including Ana Martinez, Maribel Unzueta, Ricardo Vargas, and Jesus Del Villar. Their work was made possible through generous funding from the Ballistic Missile Defense Organization, the Office of Naval Research, the Army Research Laboratory, and the National Aeronautics and Space Administration.

Our appreciation to the people at Artech House who have a done an excellent job in the production of this book. It is a pleasure to recognize the work of Jon Workman who saw that the book would be carried into production.

Finally we owe a special thanks to Dean Mensa, William Miceli, Salim Tariq, Thomas Tice, and Donald Wehner. We dully recognize their constructive critiques and sincere words of encouragement.

This book is dedicated to our families.

# 1

# Introduction

This chapter describes the problem of inverse synthetic aperture radar (ISAR) imaging and the need for motion compensation via a focal quality indicator that would measure how well target motion is estimated. This chapter also serves as a preamble to the design of advanced focal quality indicators, one of the main topics of this book.

In ISAR, the echoes of wideband signals are collected and coherently processed to generate a two-dimensional representation of a target. ISAR target images are generated by coherently processing wideband reflectivity data collected as the target rotates while remaining in the radar beam.

While the rotational motion of the target is a required element for ISAR imaging, radial and/or translational motion, henceforth defined as the target motion along the line of sight (LOS), blurs the image because it introduces phase distortion proportional to the change of the instantaneous range to the target. To generate a focused image, the kinematic parameters associated with target motion must be estimated to compensate their effect on the phase of the collected echo signal via a focal quality indicator.

A focused ISAR image can be generated through a motion compensation process using accurate target motion parameter estimates. Thus, the purpose is twofold:

- First, to design a focal quality indicator that has a unique minimum when the motion parameter estimates approach the target's actual motion parameters;
- Second, to obtain accurate motion parameters with minimal computations.

The principal topics involved with the production of high-quality images are introduced in the following five sections and developed in subsequent chapters.

## 1.1 Focal Quality Indicators

ISAR motion compensation requires knowledge of a target's instantaneous range within a range window of interest. Although the Fourier transform is the standard tool used in ISAR to measure high-resolution range walk [1], modified procedures based on adaptive joint time-frequency [2], or chirp-z processing [3] have been proposed to improve range measurement resolution. Those procedures usually measure the relative range of a prominent scatterer by tracking the peak of its envelope [4, 5], the centroid of cross-range [6], or doppler frequency shifts [2].

Once the target's instantaneous range has been calculated, target motion parameters must be determined via a focal quality indicator. Those parameters can then be used to compensate the target's ISAR signature. The motion-compensated signature is subsequently processed into focused ISAR imagery.

Focal quality indicators proposed in the literature include image entropy [7, 8], burst derivative [9, 10], phase difference [11, 12], and contrast optimization [13]. In this book, image entropy, a measure of the degree of image focusing, is used as a reference focal quality indicator. Although the entropy indicator approach is a proven and effective means for image focusing, it is an expensive proposition because it requires that the target's main scatterers be resolved in range and doppler via the two-dimensional discrete Fourier transform (DFT).

The use of the burst derivative indicator is presented as a computationally efficient alternative, because it eliminates the need of Fourier transform processing. However, the burst derivative indicator requires an advanced search method to locate the position of the global minimum among many local minima in its search space. Thus, the use of an improved focal quality indicator based on the phase slope information [14] recently has been proposed. An optimal focal quality indicator should be characterized by a global minimum and no local minima, and should avoid Fourier processing. Those characteristics would minimize the number of function evaluations by using an optimization search that leads to the best possible motion parameter estimates.

## 1.2 Radial Motion Compensation Using the Phase Method

There is an intrinsic advantage in using phase slope measurements for a target that is neither a point scatterer nor fixed in location. That advantage is discussed in [15] in the context of time-based measurements. Thus, it has been proposed to measure the instantaneous range by analyzing the phase of the target's ISAR signature sampled in the frequency domain. This book offers three approaches to measure the phase slope:

- Basic phase difference approach;
- Least squares estimation approach;
- Weighted least squares estimation approach.

We will show that the basic phase difference method avoids Fourier transform processing. However, the method is sensitive to noise. To develop a noise-resistant method, the least squares estimation is considered. That approach selects a solution closest to the true value in the least squares error sense, based on the assumption that the amplitude of a target response is stable throughout the process. However, that assumption may not be true in general. Therefore, the weighted least squares estimation is developed to account for the amplitude variations in ISAR signatures.

All three approaches exploit phase slope measurements. However, there are considerable phase errors when the amplitude is small. Complex analysis is incorporated to the weighted least squares approach to measure the phase slope more precisely. Specifically, complex analysis uses both the amplitude and the phase of the echo transfer function to select data segments, which contain accurate phase information.

## 1.3 Rotational Motion Compensation

A target that rotates slowly or is very small may not require rotational motion compensation because such a target shows small doppler frequency shift. Otherwise, the target signature requires rotational motion compensation. As in the case of radial motion compensation, rotational motion compensation can be done via a focal quality indicator. The minimum entropy method can be used to estimate motion parameters associated with a rotating target. Given a set of data collected over the integration time, the role of entropy minimization is to find the aspect change required for optimum polar reformatting. If the estimated angular motion relative to the target's center of rotation is accurate, then polar reformatting with the aspect change associated with the angular parameters followed by Fourier processing yields a focused image.

The method for finding estimates of angular velocity and acceleration is analogous to finding estimates of initial velocity and acceleration for radial motion compensation. The only difference lies in the way that the burst data are compensated. For radial motion, the exponential term due to target translational motion on the LOS is eliminated by phase compensation prior to Fourier processing. In contrast, for rotational motion, polar reformatting is used to adjust the amplitude and phase of burst data. There are several polar reformatting methods, depending on the method employed to interpolate the signatures. This book discusses the interpolation techniques of approximation, one-dimensional interpolation, and weighted integration.

## 1.4 Selective Motion Compensation Technique

This book also discusses the need for selective motion compensation. The ultimate goal of this technique is to obtain focused imagery for multiple moving-target scenarios.

> "The problem of imaging and identification may be greatly simplified by achieving selective compensation. A possible scenario can consist of surface moving vehicles where the target (along with its terrain-reflected image) could be considered to be made up of a non-rigid collection of rigid subtargets that are each moving in different, but related and somewhat predictable ways. Thus, the importance of selective compensation of typical objects on a large tactical ground vehicle might be the possibility of imaging and detecting its wheels, axles, cab body, and trailer." (Marshall Greenspan, Director of Technology, Northrop Grumman Norden Systems, email conversation, December 1996)

An approach is presented for separating and compensating wideband signatures of radar targets. The signatures thus modified are processed into focused radar imagery. In this approach, the time-frequency representation of superimposed signatures is utilized to define selective binary filters, which are enhanced and labeled using computer vision techniques. On filtering, the signature of each target is synthesized from its short-time Fourier transform representation. Following synthesis, the filtered signature is motion compensated in the frequency domain. Radar images can then be generated for scenarios in which multiple moving targets cannot otherwise be imaged. Data collected with an actual stepped frequency radar system are processed to demonstrate that combined signatures of aircraft are effectively separated, compensated, and transformed into individual ISAR images.

Figure 1.1 is an example of selective ISAR image manipulation. At the top of the figure, a fast rotating antenna of a ship causes extreme blurring. The goal

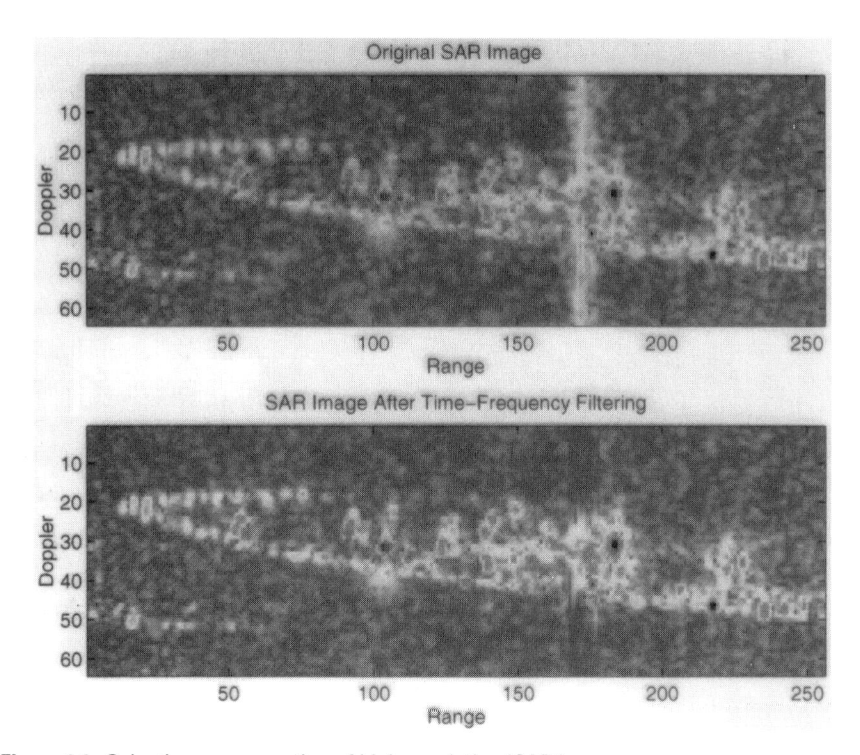


**Figure 1.1** Selective compensation of high-resolution ISAR imagery.

is to be able to manipulate parts of the image to eliminate such undesirable blurring, as shown in the lower image. No blurring caused by the antenna is visible thanks to the use of the approach proposed.

A different methodology is the use of the Gabor wavelet transform using selective motion compensation. As indicated in [16, 17], the Gabor wavelet can be used in ISAR imagery to avoid compensation. Blurring caused by rotational motion can be avoided if the target experiences only a few degrees of motion. For the translational motion, that is not the case, because the blurring deterioration of the image is more severe even at low speeds and accelerations. Thus, the selective-compensation approach can be used with the Gabor wavelet to obtain a focused image. The best optimum integration time to consider during the imaging process is formulated.

## 1.5 Sidelobe Apodization

Chapter 14 discusses a methodology that is not related to motion compensation but that does take part in the final imaging steps. The methodology can significantly improve the images obtained when it is applied in combination with one of the motion compensation techniques discuss throughout the book. This methodology involves sidelobe apodization using parametric windows. A generalization of this method is presented to obtain lower sidelobe levels and better resolution between point scatterers in an image. Thus, point scatterers reflected by a target are discernible when buried in high-intensity sidelobes.

## References

[1] Wehner, D. R., *High-Resolution Radar*, Norwood, MA: Artech House, 1995.

[2] Wang, Y., H. Ling, and V. C. Chen, "ISAR Motion Compensation via Adaptive Joint Time-Frequency Technique," *IEEE Trans. on Aerospace and Electronic Systems*, Vol. 34, No. 2, Apr. 1998, pp. 670–677.

[3] Martinez, A., *Parameter Estimation Theory for Synthetic High Resolution Radar*, Master's thesis, Univ. of Texas at El Paso, July 1998.

[4] Chen, C. C., and H. C. Andrews, "Target-Motion-Induced Radar Imaging," *IEEE Trans. on Aerospace and Electronic Systems*, Vol. AES-16, No. 1, Jan. 1980.

[5] Wu, H., et al., "Translational Motion Compensation in ISAR Image Processing," *IEEE Trans. on Image Processing*, Vol. 4, No. 11, Nov. 1995, pp. 1561–1571.

[6] Itoh, T., H. Sueda, and Y. Watanabe, "Motion Compensation for ISAR via Centroid Tracking," *IEEE Trans. on Aerospace and Electronic Systems*, Vol. 32, No. 3, July 1996, pp. 1191–1197.

[7] Flores, B. C., S. D. Cabrera, and A. Martinez, "Advances in Automatic Estimation and Compensation of Target Kinematics for Improved Radar Imaging," *Proc. SPIE Automatic Object Recognition*, Vol. 2234, 1994, pp. 49–56.

[8] Flores, B. C., *Robust Methods for the Motion Compensation of Inverse Synthetic Aperture Radar Imagery*, Ph.D. diss., Arizona State Univ., Tempe, Aug. 1990.

[9] Bocker, R. P., and S. A. Jones, "ISAR Motion Compensation Using the Burst Derivative Measure as a Focal Quality Indicator," *Internat. J. of Imaging and Technology*, Vol. 4, 1992, pp. 285–297.

[10] Flores, B. C., J. S. Son, and A. Martinez, "Motion Compensation of ISAR Imagery via the Burst Derivative Measure," *Proc. SPIE Radar/Ladar Processing and Applications*, Vol. 2562, 1995. pp. 125–136.

[11] Son, J. S., S. Tariq, and B. C. Flores, "Phase Difference Method for Target Motion Compensation of Stepped-Frequency ISAR Signature," *Proc. SPIE Radar Processing, Technology, and Applications*, Vol. 2845, 1996, pp. 163–174.

[12] Berizzi, F., and G. Corsini, "Autofocusing of Inverse Synthetic Aperture Radar Images Using Contrast Optimization," *IEEE Trans. on Aerospace and Electronic Systems*, Vol. 32, No. 3, July 1996.

[13] Son, J. S., B. C. Flores, and S. Tariq, "An Efficient Target Motion Compensation Method for Stepped Frequency ISAR Signatures," *Proc. SPIE Radar Processing, Technology, and Applications II*, Vol. 3161, 1997, pp. 20–28.

[14] Son, J. S., and B. C. Flores, "ISAR Range Measurement via Complex Analysis and Its Application to Motion Compensation," *Proc. SPIE Radar Processing, Technology, and Applications III*, 3462, 1998, pp. 44–51.

[15] Rihaczek, A. W., and S. J. Hershkowitz, *Radar Resolution and Complex-Image Analysis*, Norwood, MA: Artech House, 1996.

[16] Chen, V. C., "Reconstruction of Inverse Synthetic Aperture Radar Image Using Adaptive Time-Frequency Wavelet Transform," *Proc. SPIE Wavelet Applications for Dual Use*, Vol. 2491, 1995, pp. 373–386, 1995.

[17] Chen, V. C., and S. Qian, "Joint Time-Frequency Transform for Radar Range-Doppler Imaging," *IEEE Trans. on Aerospace and Electronic Systems*, Vol. 34, No. 2, Apr. 1998, pp. 486–499.

# 2

# ISAR Concepts

This chapter describes basic concepts of ISAR. The method for ISAR data collection is described and the effects of target motion on the phase of an ISAR target signature are illustrated. It is shown that the signature must be phase compensated to generate focused high-resolution ISAR imagery. Also, the relations of radar bandwidth and integration time to resolution are discussed. In addition, a method for image generation using the two-dimensional DFT is described.

## 2.1 ISAR Data Collection and Resolution

### 2.1.1 ISAR Geometry

ISAR is a version of synthetic aperture radar (SAR) that can be used to image targets such as ships and aircraft. Usually, a SAR image is generated from reflectivity data collected as the radar platform moves past the target area to be mapped. In contrast, an ISAR image is generated from reflectivity data collected as the target rotates while remaining in the radar beam.

The geometry in Figure 2.1 depicts an ISAR target with translational and rotational motion relative to the radar platform in a two-dimensional space. For simplicity, an aircraft moving in a plane relative to stationary radar is considered. The angle $\theta$ is the instantaneous rotational position of the target in the x-y Cartesian coordinate system, where the x-axis lies along the radar LOS. In addition, a second right-hand Cartesian coordinate system $x'$-$y'$ is assigned to the aircraft that describes the rotation of the aircraft with respect to the x-y system. Thus, the angular displacement, $\theta$, describes the rotational position of the aircraft relative to x-y. $r(0)$ is the initial range to the center of the target rotation and $r(t)$ is the instantaneous slant range at time $t$.

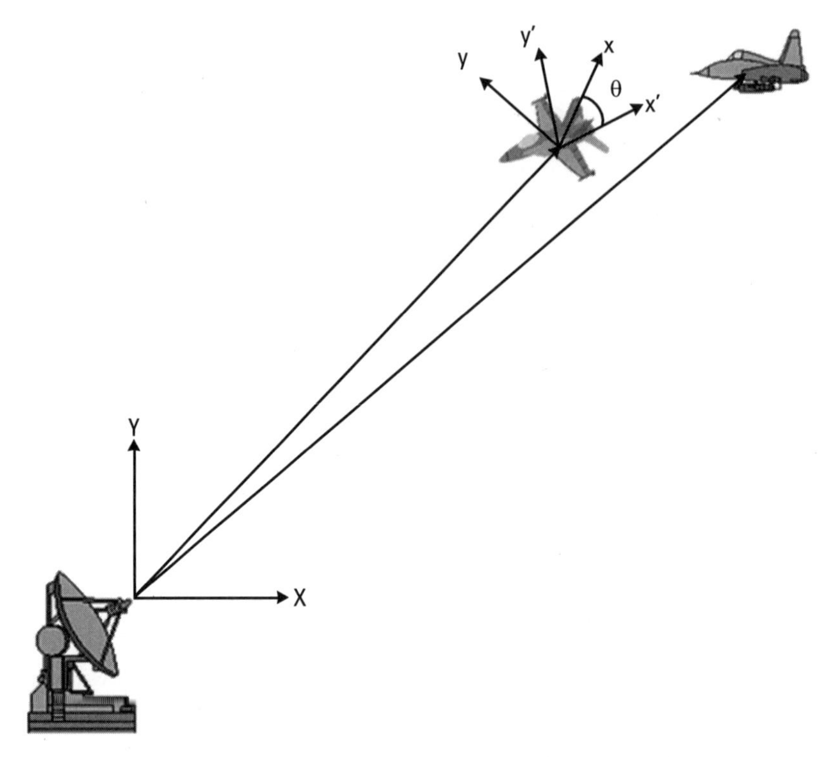


**Figure 2.1** ISAR geometry showing a moving target in a two-dimensional space.

### 2.1.2 ISAR Data Collection

A stepped-frequency waveform can be viewed as a series of $M$ bursts, where each burst consists of $N$ stepped-frequency pulses, as illustrated in Figure 2.2.

For this waveform in Figure 2.2, the frequency response of the target $U$ can be written in terms of target echo transfer function $H$ as

$$U = A \exp(i4\pi rf/c)H(\hat{p}, \hat{u}) \tag{2.1}$$

where

$$H(\hat{p}, \hat{q}) = \iint h(x', y')\exp[i2\pi(\hat{p}x' + \hat{q}y')]dx'dy' \tag{2.2}$$

Equation (2.2) is the two-dimensional Fourier transform of the target's reflectivity density function $h(x', y')$. In (2.1), $f$ is the frequency, $r$ is the instantaneous target range, and $c$ is the speed of light. $A$ is an amplitude associated with the transmitted signal, antenna gain, propagation attenuation, receiver, processing gain, and radar system loss [1]. Because $A$ is treated as a

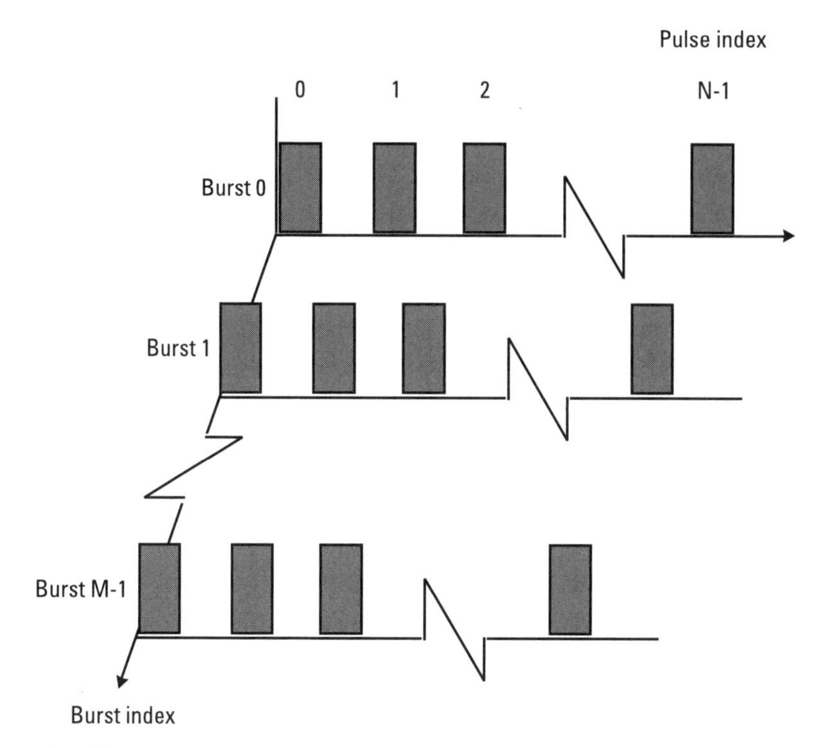


**Figure 2.2** Waveform representation of *M* bursts of *N* stepped-frequency pulses.

constant, $A$ can be set to unity without affecting any phase terms of the frequency response.

In (2.2), $\hat{p} = 2(f/c)\sin\theta$, $\hat{q} = 2(f/c)\cos\theta$, where $\theta$ is the angle between the radar axis and target axis. For stepped-frequency ISAR signatures, the frequency $f$ takes on the value $f_n$, while $r$ becomes $(r_0 + r_{m,n})$ and $\theta$ becomes $\theta_{m,n}$. Here, $r_0$ denotes the initial target range and $r_{m,n}$ is the time-dependent range component. Consequently, $\hat{p}$ and $\hat{q}$ become $\hat{p}_{m,n} = 2(f_n/c)\sin\theta_{m,n}$ and $\hat{q}_{m,n} = 2(f_n/c)\sin\theta_{m,n}$, respectively, where $m$ and $n$ are the corresponding burst and pulse indices.

The frequency response $U$ can be written as

$$U(m, n) = \exp[i4\pi(r_0 + r_{m,n})]H(m, n) \tag{2.3}$$

where $H(m, n)$ is defined as

$$H(m, n) = \iint h(x', y') \exp[i2\pi(\hat{p}_{m,n}x' + \hat{q}_{m,n}y')]dx'dy' \tag{2.4}$$

A second-order parametric model for $r_{m,n}$ would be

$$r_{m,n} = v_0 t_{m,n} + (\tfrac{1}{2}) a_0 t_{m,n}^2 \tag{2.5}$$

Notations denoted by $v_0$ and $a_0$ in (2.5) are the initial velocity and the acceleration of the target centroid, respectively. The sampling time, $t_{m,n}$, referenced to the initial moment of sampling is given by

$$t_{m,n} = (n + mN)\Delta t \tag{2.6}$$

where $\Delta t$ is the pulse repetition interval (PRI). In (2.3), the frequency of the $n$th pulse in a burst is

$$f_n = f_0 + n\Delta f \tag{2.7}$$

where $f_0$ is the initial frequency in a burst and $\Delta f$ is the frequency step from pulse to pulse.

### 2.1.3 ISAR Slant Range and Cross-Range Resolution

The stepped-frequency waveform relaxes wide instantaneous bandwidth and high sampling rate requirements by sampling a near-steady-state reflectivity version of $H(\hat{p}, \hat{q})$ at discrete frequencies stepped pulse to pulse. High-range resolution is achieved by Fourier processing the waveform that synthetically spans a bandwidth. The fundamental relationship for the range resolution and waveform bandwidth is expressed as

$$\Delta r = c/(2\beta) \tag{2.8}$$

For $N$ pulses stepped in frequency by an amount $f$, the slant range resolution is

$$\Delta r = c/[2(N - 1)\Delta f] \tag{2.9}$$

Cross-range resolution, $\Delta r_c$, depends on the resolvable difference in the doppler frequencies of two scatterers in the same slant-range cell. Furthermore, doppler resolution is related to the available coherent integration time $T$. The integration time is almost the same as the time required to collect $M$ frequency responses. Because the reflectivity samples of the same frequencies are taken every $N\Delta t$ seconds, the doppler resolution is given by

$$\Delta f_D = 1/[N(M-1)\Delta t] \approx 1/T \tag{2.10}$$

Therefore, the cross-range resolution is written as [1]

$$\Delta r_c = c\Delta f_D/(2\omega_0 f_c) = \lambda/(2\omega_0 T) \tag{2.11}$$

where $\omega_0$ is the rotation rate of the target and $\lambda = c/f_c$ is the center wavelength of the waveform.

## 2.2 Stepped-Frequency ISAR Processing

### 2.2.1 ISAR Image Formation

Figure 2.3 illustrates the image-processing sequence for an ISAR data matrix made of $M$ bursts (rows) of $N$ complex echo samples (columns). Correction for target translational motion is assumed to have been made. Let $H(k, j)$ be the response at frequency $f_j$ for burst $k$. Obviously, $H$ is a two-dimensional sequence of size $(M \times N)$.

To obtain the range profile history from $H(k, j)$, the inverse discrete Fourier transform (IDFT) of each of the $M$ bursts of $N$ frequency samples is taken, row by row, to generate $M$ new rows of synthetic range profiles, each with $N$ synthetic range cells. Denote that profile history as $h(k, n)$, where $n$ is the range cell index. The basic operation that transforms frequency data into range-reflectivity is

$$h(k, n) = 1/N \sum_{j=0}^{N-1} H(k, j) \exp(i2\pi nj/N) \tag{2.12}$$

for $n = 0, 1, \ldots, N-1$. Scatterers are now resolved in slant range.

The time-history matrix in (2.12) can be transformed into a third matrix by taking the DFT of the $M$ complex values in each range cell, column by column. That is, each record of the complex reflectivity in slant range cell is transformed into a reflectivity profile in the cross-range via the DFT. Denote the output of the transform as $D(m, n)$, where $m$ is the cross-range cell index. Hence,

$$D(m, n) = 1/M \sum_{k=0}^{M-1} h(k, n) \exp(-i2\pi km/M) \tag{2.13}$$

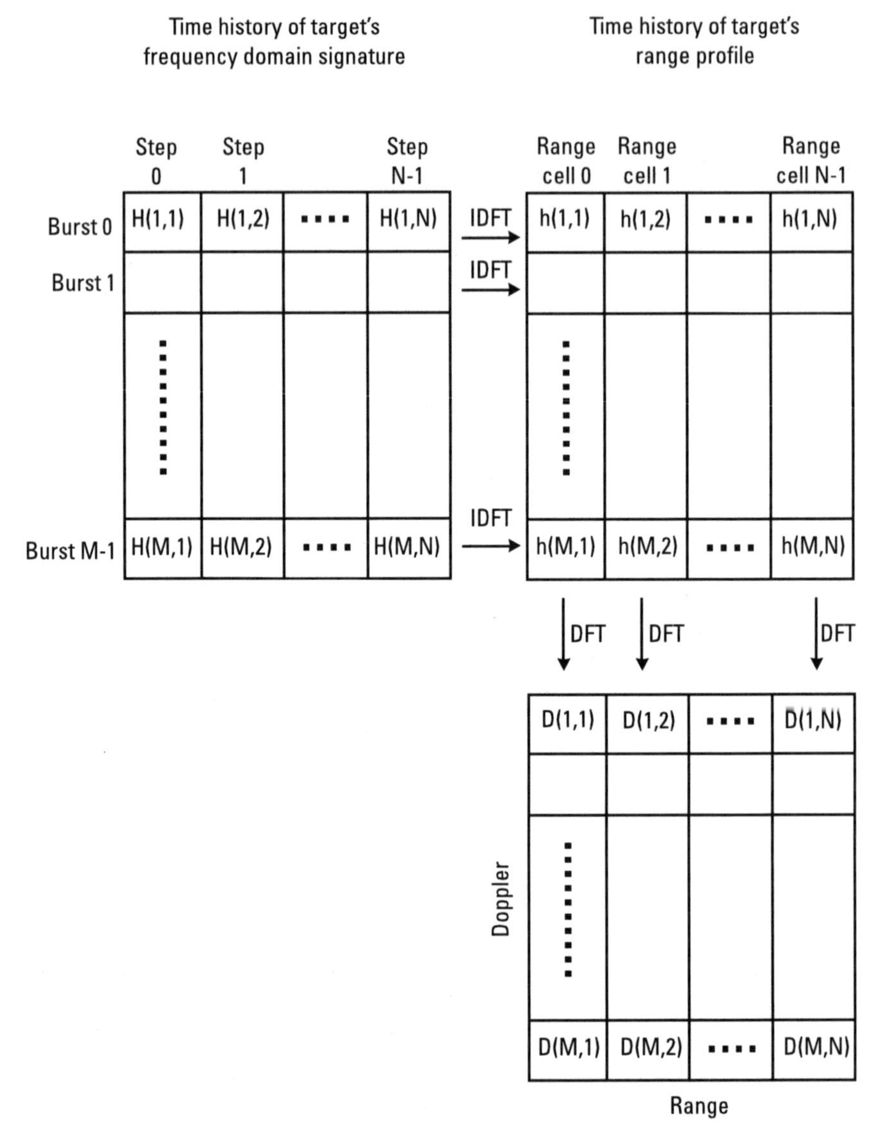


**Figure 2.3** Row-column decomposition for synthetic ISAR imaging.

for $m = 0, 1, \ldots, M - 1$. The result after converting complex values of $D(m, n)$ to absolute values is a range-doppler matrix, which is the ISAR image.

From the preceding discussion, it is clear that ISAR image formation entails a row-column decomposition of the two-dimensional Fourier transform. Writing the process of (2.12) and (2.13) in compact form, the two-dimensional transform is expressed as

$$D(m, n) = 1/(MN) \sum_{k=0}^{M-1} \sum_{j=0}^{N-1} H(k, j) \exp(i2\pi nj/N) \exp(-i2\pi km/M) \tag{2.14}$$

for $m = 0, 1, \ldots, M - 1$ and $n = 0, 1, \ldots, N - 1$.

### 2.2.2 Range Walk and Range Offset

Target translational motion produces both profile-to-profile range walk and a constant range offset. Range offset is a fixed offset for all profiles due to constant radial velocity. Thus, range offset does not produce much distortion in ISAR imagery. In contrast, range walk accumulates from profile to profile so that range profiles may end up misaligned for a moving target. Figure 2.4 shows a typical case of range walk and range offset. This figure is a range profile history of a Boeing 727.

For large $M$, the accumulated number of range cells shifted due to target velocity during the entire image frame time is

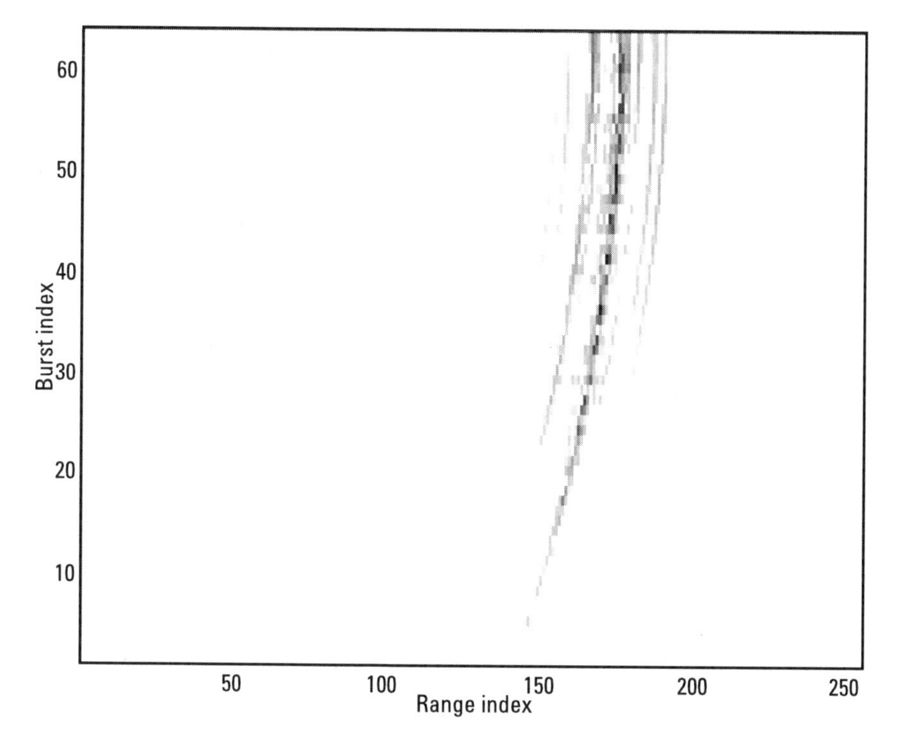


**Figure 2.4** Range profile history of Boeing 727.

$$N_{walk} \approx -2n\Delta t/c(f_0 + NM\Delta f)v_0 = -(2N\Delta t f_0 + 2MN\Delta t n\Delta f)v_0/c \tag{2.15}$$

Here, the range offset is given by the first term in (2.15), while the range walk is given by its second term.

Uncorrected range walk affects ISAR imagery. Responses from scatterers walk out of the synthetic range cell columns in which the IDFT is performed to resolve scatterers in the cross-range. This range walk degrades cross-range resolution since its net effect is to reduce the integration of the scatterer's response in time. Moreover, scatterer responses walk into adjacent range-cell columns, producing range spread. Figure 2.5 shows the image of a Boeing 727 aircraft obtained by processing uncompensated ISAR data. The figure fails to show a discernable aircraft shape. From this example in Figure 2.5, it is clear that motion compensation is a requirement for obtaining high-resolution radar imagery.

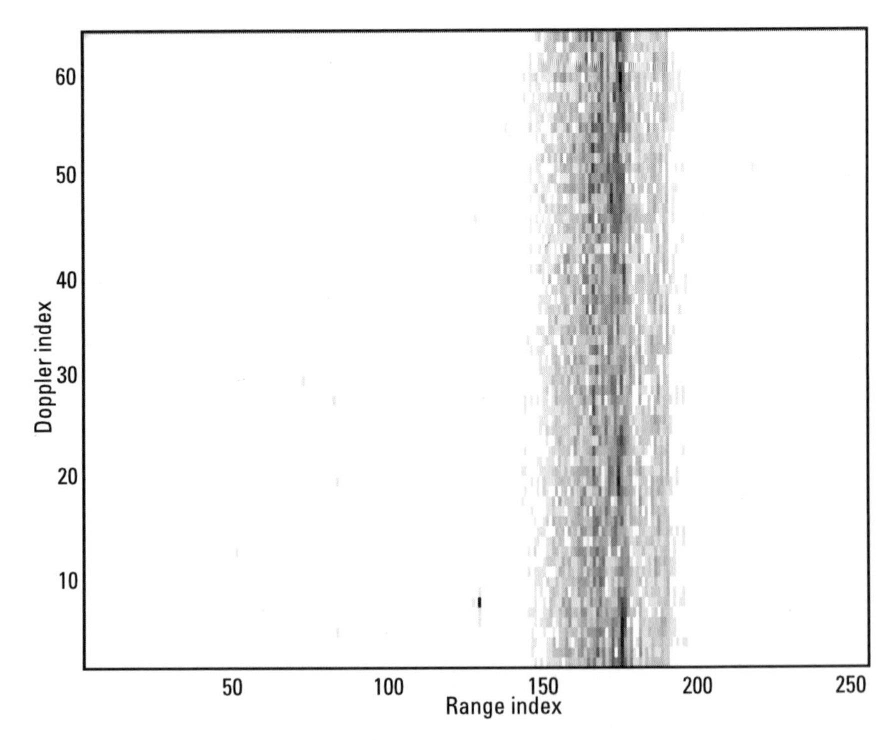


**Figure 2.5** Image of uncompensated Boeing 727.

### 2.2.3 Motion Compensation

Translational motion compensation algorithms estimate $H(m, n)$ from the unprocessed ISAR spatial frequency spectrum, $U(m, n)$. That process requires the estimation of target motion parameters so that most of the phase errors introduced by the exponential term in (2.3) are eliminated. Assume that $v$ and $a$ are the velocity and acceleration estimates of $v_0$ and $a_0$, respectively. Performing motion compensation on the received echo transfer function is tantamount to multiplying $U(m, n)$ by a factor whose phase is proportional to

$$vt_{m,n} + \frac{1}{2} at_{m,n}^2 \tag{2.16}$$

The motion compensated signal is denoted by $\tilde{H}$, and it is an estimate of $H(m, n)$. $\tilde{H}$ is then given by

$$\tilde{H}(m, n) = \exp[-i4\pi f_n/c(vt_{m,n} + \frac{1}{2} at_{m,n}^2)]\, U(m, n) \tag{2.17}$$

Substituting (2.3) into (2.17), we obtain

$$\tilde{H}(m, n) = \exp[-i4\pi f_n/c(r_0 + r_{m,n})$$
$$\times \exp[-i4\pi f_n/c(vt_{m,n} + \frac{1}{2} at_{m,n}^2)] H(m, n) \tag{2.18}$$

Further simplification of (2.18) leads to

$$\tilde{H}(m, n) = \exp[-i4\pi f_n/c(r_0 + r_{m,n})] H(m, n) \tag{2.19}$$

where

$$\Delta r_{m,n} = \Delta vt_{m,n} + \frac{1}{2} \Delta at_{m,n}^2 \tag{2.20}$$

Notice that $\Delta v = v_0 - v$ and $\Delta a = a_0 - a$ are parameter estimation errors. When there is no parameter estimation error (i.e., $\Delta v = 0$, $\Delta a = 0$), then $\Delta r_{m,n}$ is zero. Hence, the exponential term in (2.19) becomes a constant, and $\tilde{H}(m, n)$ becomes the best approximation of $H(m, n)$.

Figures 2.6 and 2.7 are the range profile history and range-doppler image of a motion-compensated Boeing 727 signature. If we compare those figures to Figures 2.4 and 2.5, the effect of motion compensation in ISAR is evident.

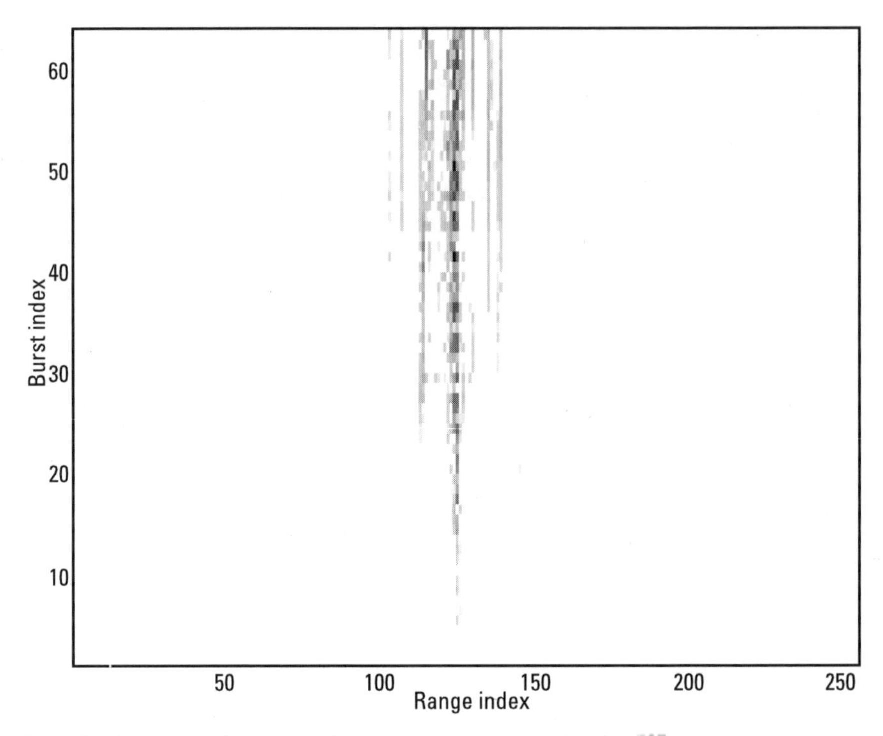


**Figure 2.6** Range profile history of a motion-compensated Boeing 727.

### 2.2.4 ISAR Imaging of a Point Target

An ideal point target is used to simulate the effects of target velocity in ISAR imagery. Figure 2.8 is an ISAR image of the point target rotating about the center of the frame with an angular velocity of 0.02 rad/s. The point target is located at point (10,10) ($m$). The aspect change is 0.0607 (rad). Using (2.9) and (2.11), the slant range resolution is calculated to be 0.5421 ($m$), and the cross-range resolution is 0.7871 ($m$). The image shows the peak response at slant range $r_s = 10$ ($m$), and the cross-range $r_c = 10$ ($m$) in agreement with the initially given position values.

Figure 2.9 is an image of the same point target except that the target has a translational velocity of $v_0 = 5$ m/s. The resulting image is blurred and we cannot locate the peak position precisely. Using the second term (2.15), the range walk is calculated to be 28 range-cells. Multiplying that number by the slant range and cross-range resolution, a slant range spread of 15.19 ($m$) and a cross-range spread of 22.04 ($m$) are estimated.

Figure 2.10 is an image of a target with $v_0$ of 10 m/s. The image is spread over twice the range and cross-range of the image in Figure 2.9. Because the

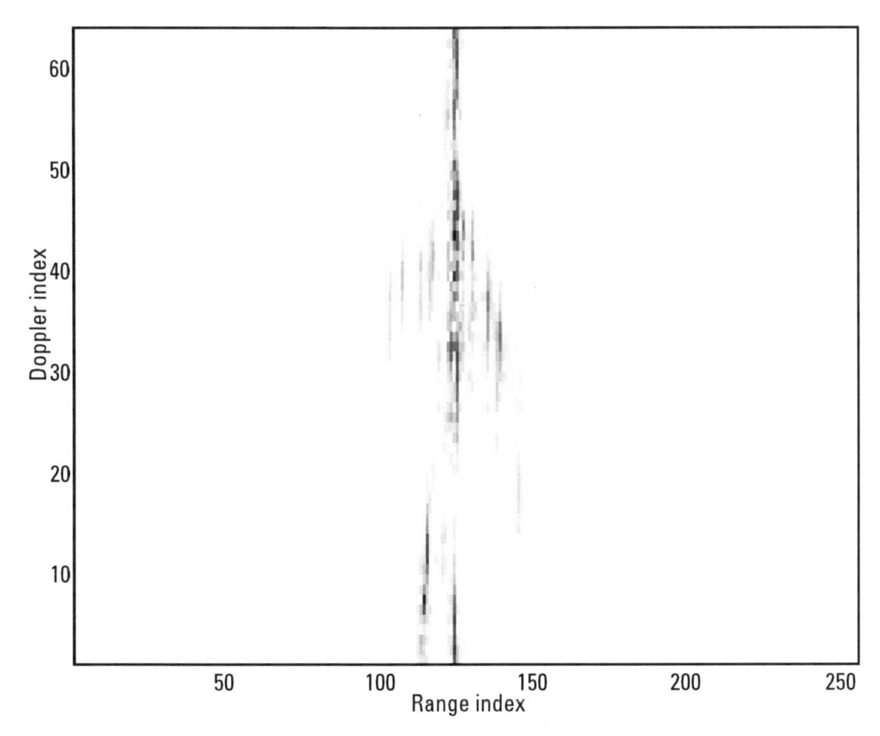


**Figure 2.7** Image of motion-compensated Boeing 727.

range-cell walk using the second term (2.15) is calculated to be 56 range-cells, the corresponding slant range spread is 30.38m and cross range spread is 44.08m.

When the signal is properly motion compensated with its respective actual motion parameters, the resulting image is identical to Figure 2.8, which does not suffer any translational motion. Figures 2.8, 2.9, and 2.10 show that motion compensation is an indispensable procedure to generate a focused ISAR imagery even for the simplest scatterers.

## 2.3 Entropy Measurement

By definition, a focal quality indicator indicates how well the motion estimation process is performed so that the motion-compensated target image is focused. Image entropy is a traditional focal quality indicator used for ISAR motion compensation. This indicator is based on the notion that an image is perfectly focused when its entropy is minimized. Throughout the discussion of various motion compensation approaches, the entropy indicator is used for reference.

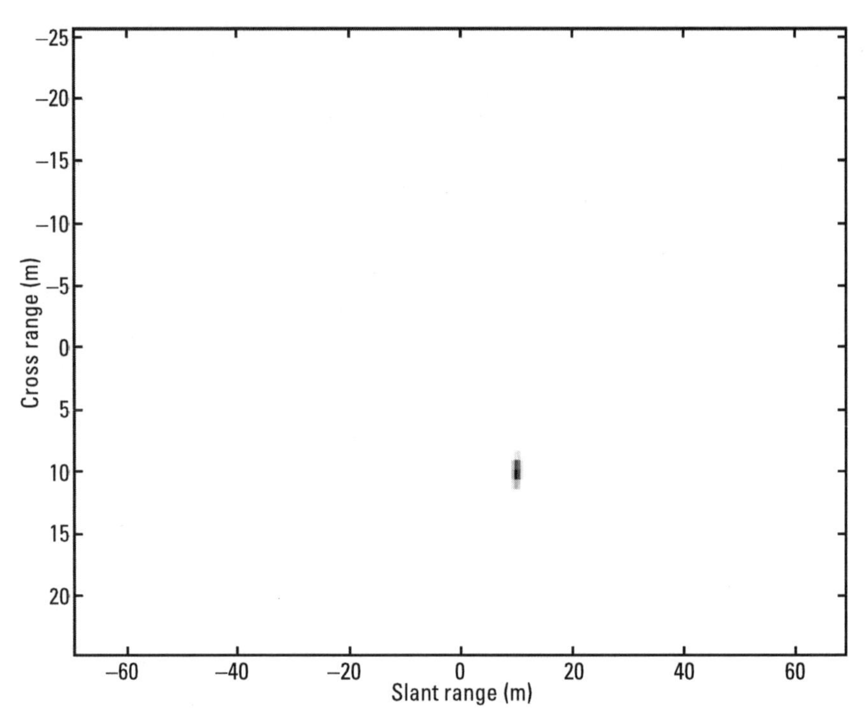


**Figure 2.8** Image of a point target with translational velocity $v_0 = 0$ (m/s).

### 2.3.1 Definition of Entropy

Let $x$ be a real random variable with the probability density function (PDF) $p(x)$. Shannon [2] defined the entropy of $x$ as

$$S = -\int_{-\infty}^{\infty} p(x) \ln p(x) dx \tag{2.21}$$

Following Shannon's definition of entropy, Pun [3] defined the entropy of the image as

$$S = -\sum_{0}^{MN-1} P_k \ln P_k; \quad P_k = z_k = z_k / z \tag{2.22}$$

where $z_k$ is the frequency of the gray-level $k$ in an image. $z$ is defined as

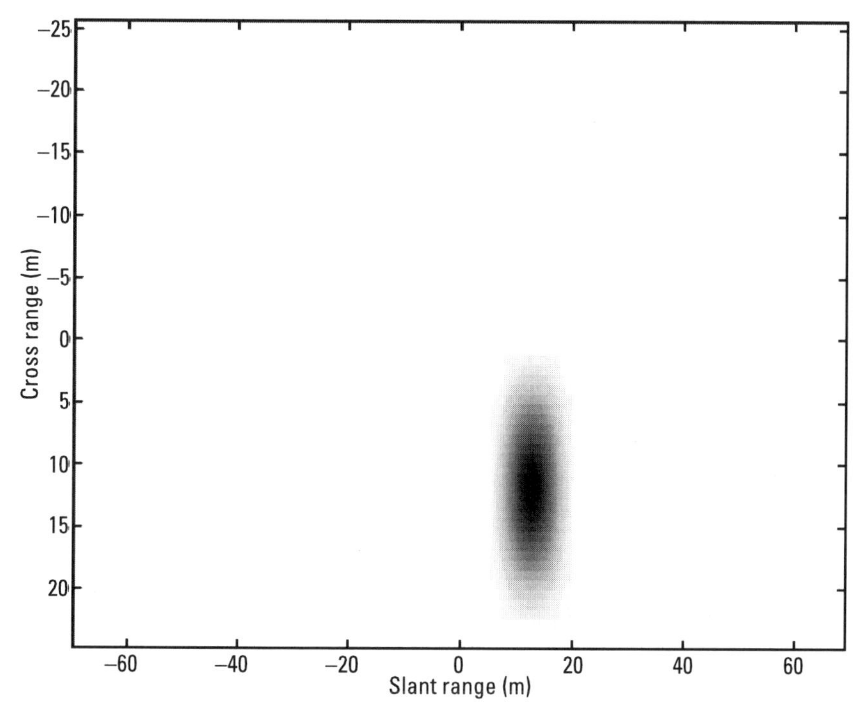


**Figure 2.9** Image of a point target with translational velocity $v_0 = 5$ m/s.

$$z = \sum_{0}^{MN-1} P_k \tag{2.23}$$

It follows that $p_k$ is a function of pixel intensity in an image. The distribution of pixel intensity is commensurate with the degree of image focus. In that context, the entropy is a measure of probabilistic uncertainty in information. The uncertainty increases as the image becomes more blurred and decreases as the image becomes more focused. Therefore, we state that the image is best focused when the entropy is minimum.

Suppose the random variable $x$ is normally distributed. If $x$ is to represent the position of a randomly moving particle, $p(x)$ is the PDF of its position. It is easy to show that its entropy is given by

$$S = \ln(\sigma\sqrt{2\pi e}) \tag{2.24}$$

Notice that the entropy of $x$ is proportional to its standard deviation $\sigma$. A small value for $\sigma$ implies that the probability of finding the particle is large in

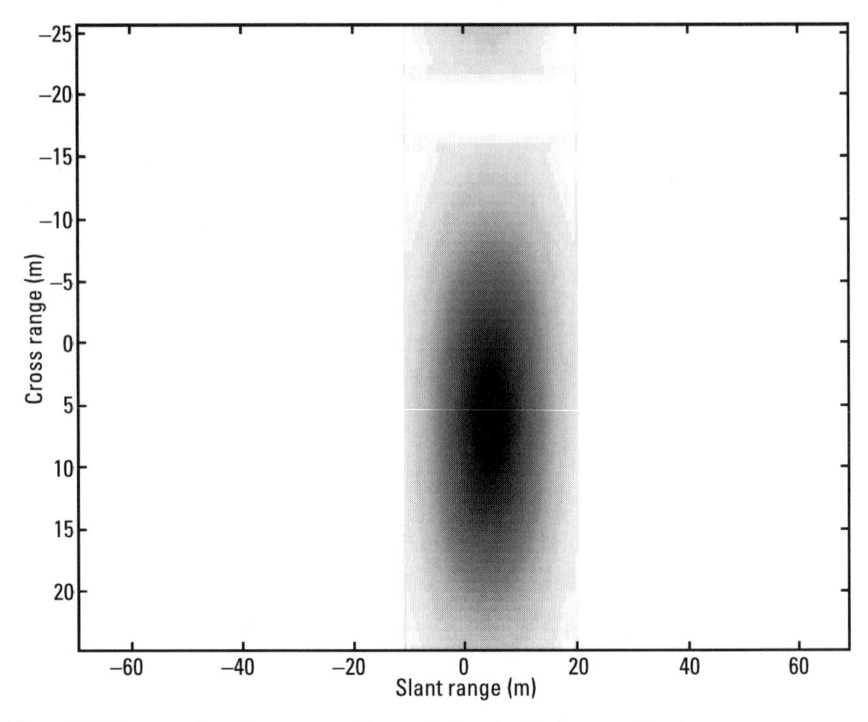


**Figure 2.10** Image of a point target with translational velocity $v_0 = 10$ m/s.

a relatively narrow space and the entropy would be small. An increase in the standard deviation yields a larger value of the entropy and would imply that the particle can be found at a position with less accuracy. Figure 2.11 shows the behavior of (2.24). In our work, the normalized intensity of an image $I_{m,n}$ is interpreted as a version of the probability density function of a two-dimensional random process whose entropy is to be minimized.

### 2.3.2 Image Entropy

The entropy concept discussed in Section 2.3.1 has been applied to the motion compensation of target imagery obtained via ISAR. In effect, the image entropy is viewed as a measure of the target's image quality. The target can be assumed to be a collection of moving point scatterers that reflect electromagnetic plane wave while immersed in the radar beam. Using samples of the echo transfer function, the positions of those scatterers have to be found by Fourier processing to construct a clean target image. In practice, the energy scattered from a given point scatterer is not localized but spread over several resolution cells due to target motion. Thus, it is intuitive to regard that power distribution as an

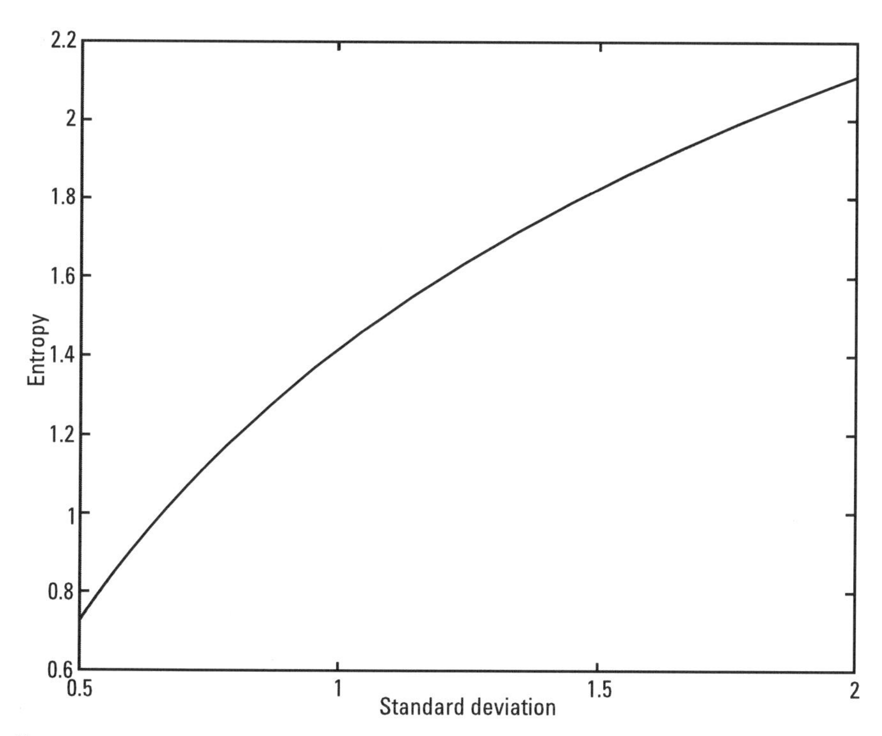


**Figure 2.11** Entropy of Gaussian random variable as a function of standard deviation.

image density function. Poor phase compensation corresponds to a high standard deviation in range accuracy, resulting in a high entropy. Consequently, the image is in an unfocused state. Thus, higher entropy would mean lower image accuracy, while lower entropy would imply higher accuracy. As the target motion is compensated with improved motion parameter estimates, the entropy decreases to a minimum.

Equation (2.22) can be rewritten as

$$S = -\Sigma\Sigma I_{m,n} \ln I_{m,n} \tag{2.25}$$

where

$$I_{m,n} = \tilde{D}_{m,n} / \Sigma\Sigma\, \tilde{D}_{m,n} \tag{2.26}$$

and $\tilde{D}_{m,n}$ is the high-resolution radar cross-section of an individual image element given by (2.13) or (2.14). Notice that $\tilde{D}_{m,n}$ is the Fourier transform of the compensated reflectivity density function $\tilde{H}_{m,n}$.

A variation of (2.25) is given by an exponential information gain [4] as

$$S = \Sigma\Sigma I_{m,n} \exp(1 - I_{m,n}) \tag{2.27}$$

That expression tends to be less susceptible to noise.

The sharpness of the image depends on the estimates of target kinematic parameters. The compensation process is repeated for different sets of velocity and acceleration estimates until a translational motion solution is found that yields a global minimum entropy value. It should be pointed out that this method is computationally expensive because it requires a two-dimensional DFT for each image generated, and the image generation process has to continue until a global minimum entropy value is reached.

### 2.3.3 Results

Figure 2.12 shows the results of an enumerative search using the image entropy defined by (2.27) for a point target with infinite signal-to-noise ratio (SNR). In this case, the actual target motion coefficients are set to (−31.5 m/s, 3 m/s²). The most remarkable feature of the resulting entropy surface is its convexity

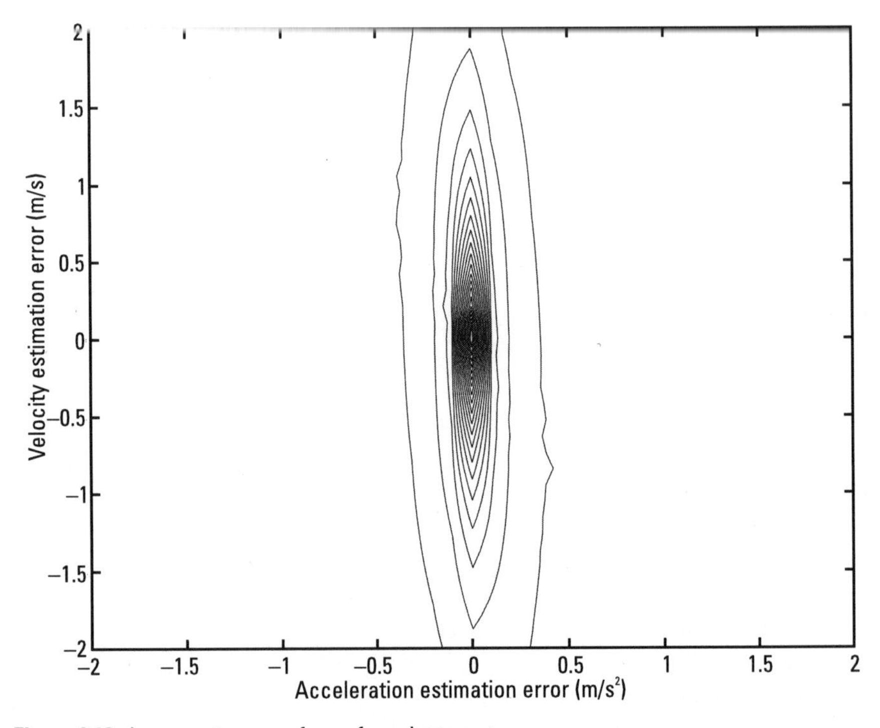


**Figure 2.12** Image entropy surface of a point target.

near the center of the search space where the actual target motion coefficients reside. Outside that area, the entropy surface is almost flat. Optimization algorithms have difficulties in finding the global minimum when we have no prior knowledge about the target motion parameters.

Figure 2.13 shows the results of an enumerative search using the image entropy defined by (2.27) applied to a set of Boeing 727 data. Even though the entropy surface shows a unique solution, some local minima show up in the figure. This behavior is anticipated for man-made targets that cannot be modeled as a set of point targets or when the SNR is low.

## 2.4 Summary

While rotational motion is a required element in ISAR imaging, translational motion blurs ISAR imagery. That blurring effect is more notorious for high-range-resolution radar.

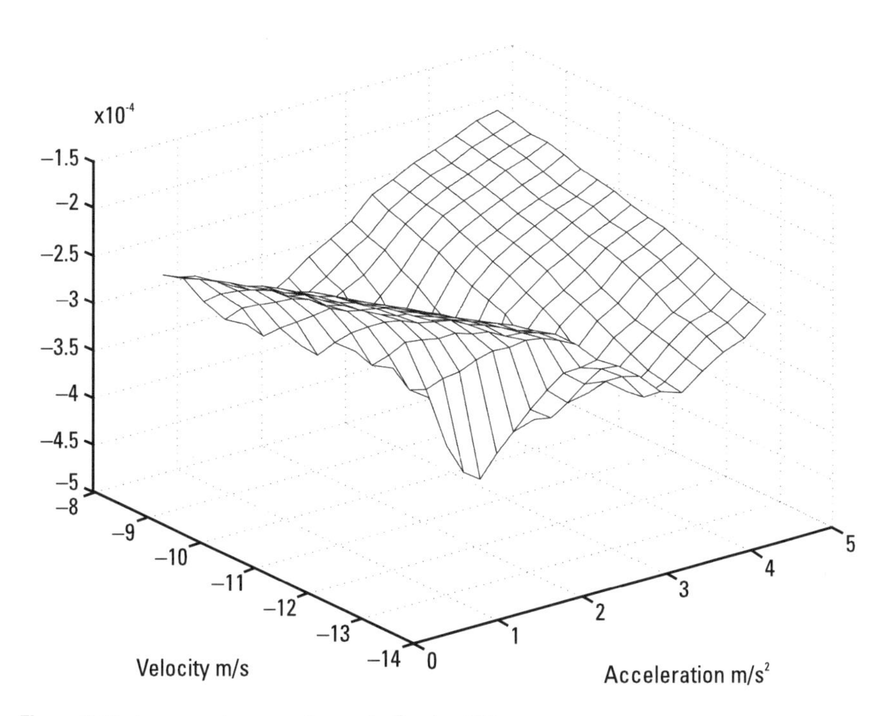


**Figure 2.13** Image entropy surface of a Boeing 727.

To focus a blurred image, the effect that translational motion has on the phase of the ISAR signature must be reduced through motion compensation prior to ISAR image formation. Motion compensation requires knowledge of the target's kinematic motion. Accurate target motion estimation is possible using the entropy-based iterative approach. If the initial guess of motion parameters is close to the actual motion parameters, a solution is guaranteed. When a reasonable initial guess is not available, the number of iterations required to focus the image may be considerable. The outstanding issues of the entropy motion compensation method can be summarized as follows:

- The method is based on nondiscriminatory data processing.
- It requires intensive computational effort.
- It is susceptible to noise.

## References

[1] Wehner, D. R., *High-Resolution Radar*, Norwood, MA: Artech House, 1995.

[2] Shannon, C. E., "A Mathematical Theory of Communication," *Bell Syst. Tech. J.*, Vol. 27, July 1948, pp. 379–423.

[3] Pun, T., "A New Method for Gray-Level Picture Thresholding Using the Entropy of the Histogram," *Signal Processing*, Vol. 2, 1980, pp. 223–237.

[4] Nikhil, P. R., and P. K. Sankar, "Entropy: A New Definition and Its Applications," *IEEE Trans. on Systems, Man and Cybernetics*, Vol. 21, No. 5, Sept. 1991, pp. 1260–1270.

# 3

# Burst Derivative Approach

The minimum image entropy approach discussed in Chapter 2 is a computationally demanding motion compensation technique because a considerable number of two-dimensional DFTs are required to generate an entropy surface over a search plane from which suitable motion parameter estimates are chosen. The burst derivative technique, originally proposed by Bocker and Jones [1], is less computationally intensive because it bypasses the need for Fourier transform computations.

This chapter formalizes the burst derivative concept, discusses the properties of the burst derivative approach, and establishes the connection between the burst derivative and the Fisher information. In addition, it shows that the burst derivative is strongly dependent on waveform parameters such as pulse repetition frequency (PRF), bandwidth, and frequency step. Those waveform parameters affect the size of the search space over which motion parameter estimation is performed.

## 3.1 The Cramer-Rao Lower Bound and Fisher Information

According to the Cramer-Rao lower bound, the mean square error (MSE) corresponding to the estimator of a parameter cannot be smaller than a certain quantity related to the likelihood function. The proof is shown in [2] and is summarized here.

Let $\hat{x}(z)$ be an unbiased estimate of the nonrandom real-valued parameter $x$ based on the observation denoted as $z$. The likelihood function of $x$ is

$$\Lambda(x) = p(z \mid x) \tag{3.1}$$

It is assumed that (3.1) has first and second derivatives with respect to $x$ and that its integral also exists. From the unbiased condition of (3.1), the estimator is written as

$$E[\hat{x}(z) - x] = \int_{-\infty}^{\infty} [\hat{x}(z) - x] p(z|x) dz = 0 \tag{3.2}$$

Taking the derivative of (3.2) with respect to $x$, we have

$$\begin{aligned} \frac{d}{dx}\int_{-\infty}^{\infty} [\hat{x}(z) - x] p(z|x) dz &= \int_{-\infty}^{\infty} \frac{\partial}{\partial x}\{[\hat{x}(z) - x] p(z|x)\} dz \\ &= -\int_{-\infty}^{\infty} p(z|x) dz + \int_{-\infty}^{\infty} [\hat{x}(z) - x] \frac{\partial p(z|x)}{\partial x} dz = 0 \end{aligned} \tag{3.3}$$

Using the identity

$$\frac{\partial p(z|x)}{\partial x} = \frac{\partial \ln p(z|x)}{\partial x} p(z|x) \tag{3.4}$$

and the fact that the first integral of the last line in (3.3) is unity yields

$$\int_{-\infty}^{\infty} [\hat{x}(z) - x] \frac{\partial \ln p(z|x)}{\partial x} p(z|x) dz = 1 \tag{3.5}$$

To apply the Schwarz inequality to (3.6), the equation is arranged as

$$\int_{-\infty}^{\infty} \{[\hat{x}(z) - x]\sqrt{p(z|x)}\}\left\{\frac{\partial \ln p(z|x)}{\partial x}\sqrt{p(z|x)}\right\} dz = 1 \tag{3.6}$$

The Schwarz inequality is

$$\int_{-\infty}^{\infty} f_1(z) f_2(z) dz \leq \|f_1\| \|f_2\| \tag{3.7}$$

where

$$\|f_i\| = \left\{\int_{-\infty}^{\infty} f_i(z)^2 dz\right\}^{1/2} \quad (3.8)$$

Substituting (3.6) and (3.8) in (3.7),

$$\left\{\int_{-\infty}^{\infty} [\hat{x}(z) - x]^2 p(z \mid x)\, dz\right\}\left\{\int_{-\infty}^{\infty} \left[\frac{\partial \ln p(z \mid x)}{\partial x}\right] p(z \mid x)\, dz\right\}^{1/2} \geq 1 \quad (3.9)$$

which is rewritten as

$$E\{[\hat{x}(z) - x]^2\} \geq \left\{E\left[\frac{\partial \ln p(z \mid x)}{\partial x}\right]^2\right\}^{-1} \quad (3.10)$$

According to [2], the Fisher information is defined as

$$FI = E\left\{\left[\frac{\partial \ln p(z \mid x)}{\partial x}\right]^2\right\} \quad (3.11)$$

Substituting (3.11) to (3.10), the Cramer-Rao lower bound is limited by the Fisher information as

$$E\{[\hat{x}(z) - x]^2\} \geq FI^{-1} \quad (3.12)$$

## 3.2 Fisher Information as a Focal Quality Indicator

In this section, a different form of Fisher information is derived for motion estimation from the definition of Fisher information given in (3.11). This equation is rewritten as

$$FI = \int_{-\infty}^{\infty} \left[\frac{\partial \ln p(z \mid x)}{\partial x}\right]^2 p(z \mid x) dz \quad (3.13)$$

Squaring the identity function (3.4) and dividing it by $p(z \mid x)$ yields

$$\left[\frac{\partial \ln p(z \mid x)}{\partial x}\right]^2 p(z \mid x) = \left[\frac{\partial p(z \mid x)}{\partial x}\right]^2 \frac{1}{p(z \mid x)} \quad (3.14)$$

Substitution of (3.14) to (3.13) yields

$$FI = \int_{-\infty}^{\infty} \left[ \frac{\partial p(z|x)}{\partial x} \right]^2 \frac{1}{p(z|x)} dz \tag{3.15}$$

Assuming $z$ and $x$ are independent, the conditional probability function is written as

$$p(z|x) = p(z) \tag{3.16}$$

Then the Fisher information of a random variable $z$ with PDF $p(z)$ is defined as

$$FI = \int_{-\infty}^{\infty} [dp(z)/d(z)]/[p(z)]d(z) \tag{3.17}$$

The Fisher information measures the degree of disorder in information. As an image becomes increasingly blurred, the disorder in information also increases. Thus, an image is best focused when the disorder in the image information is at its minimum.

If $z$ is a Gaussian random variable with zero mean and standard deviation of $\sigma$, its Fisher information is

$$FI = 1/\sigma^2 \tag{3.18}$$

Clearly, the Fisher information is inversely proportional to the squared value of the standard deviation of the random variable. Such behavior is opposite to that of the entropy. However, the negative of the Fisher information and the entropy behave in identical fashion.

Fisher information is plotted in Figure 3.1 as a function of the standard deviation for a given Gaussian random variable. Just as in the case of entropy, Fisher information value increases as the standard deviation increases. Frieden presented an analysis and discussion on the equivalence of the Fisher information and the entropy values [3].

## 3.3 Fisher Information for Motion Compensation

In the context of ISAR, the Fisher information and entropy provide similar information about the quality of an image. Whereas the entropy concept can be

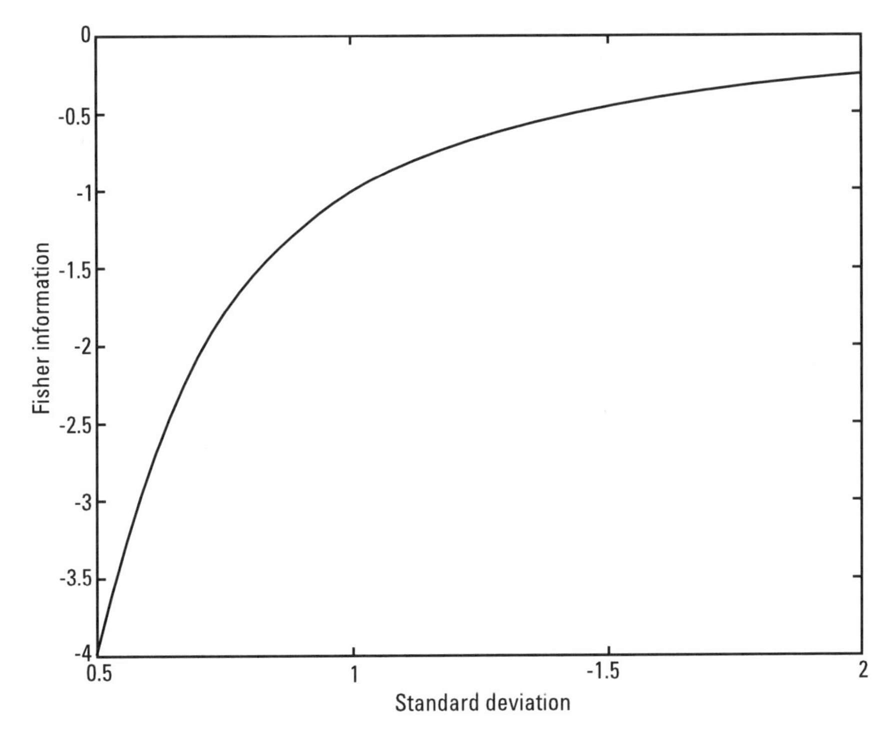


**Figure 3.1** Fisher information of Gaussian random variable as a function of standard deviation.

applied to the image of a target to measure the contrast, the Fisher information concept can be applied in the frequency domain to minimize the distortion effects of target motion. In analogy to (3.17), let us define the Fisher information for the compensated target reflectivity $\tilde{H}(m, n)$ as

$$FI = \sum_{m=0}^{M-1} \sum_{n=0}^{N-1} 1/\tilde{H}(m, n)\left|\partial \tilde{H}(m, n)/\partial m\right|^2 \tag{3.19}$$

For a point scatterer, an expression can be obtained in terms of the error in the velocity estimate:

$$FI = \sum_{m=0}^{M-1} \sum_{n=0}^{N-1} \left|4\pi N \Delta v \Delta t f_n / c\right| \tag{3.20}$$

Notice that the Fisher information reaches a minimum as the error in velocity estimate, $\Delta v$, reduces to zero.

Given that the target reflection data is in discrete form, an approximation of the Fisher information is

$$FI = \sum_{m=0}^{M-1} \sum_{n=0}^{N-1} \left| \tilde{H}(m+1, n) - \tilde{H}(m, n) \right|^2 / \left| \tilde{H}(m, n) \right| \quad (3.21)$$

Here, it is assumed that the estimated kinematic errors are sufficiently small so that phase foldover does not occur from burst to burst. Recall that $\left| \tilde{H}(m, n) \right| = 1$ for a point target with a normalized radar cross-section. Hence,

$$FI = \sum_{m=0}^{M-2} \sum_{n=0}^{N-1} \left| \tilde{H}(m+1, n) - \tilde{H}(m, n) \right|^2 \quad (3.22)$$

That expression is similar to the heuristic definition of the burst derivative given by Bocker and Jones [1]:

$$B = \sum_{m=0}^{M-2} \sum_{n=0}^{N-1} \left| \tilde{H}(m+1, n) - \tilde{H}(m, n) \right| \quad (3.23)$$

The similarity shows that a heuristically developed burst derivative also measures the sharpness of the image as Fisher information. Their experiments have shown that the burst derivative, as defined in (3.23), can be successfully used to estimate target motion parameters within a limited search space. Equation (3.22) makes it clear why (3.23) should work as a focal quality indicator in ISAR motion compensation.

## 3.4 Burst Derivative as a Function of Motion Estimates

The burst derivative was developed by the Naval Command, Control, and Ocean Surveillance Center (NCCOSC) as an empirical focal quality indicator to enhance the computational efficiency for ISAR motion compensation. Many motion compensation algorithms have been developed to remove the unwanted effects of target motion. However, many of those methods require the use of a space domain focal quality indicator such as entropy to determine image sharpness. Usually focal quality indicators are computed by processing a large number of two-dimensional fast Fourier transforms. This is due to the fact that the actual processing of ISAR data is performed in the spatial frequency domain and not in the space domain where the final ISAR image is displayed. The burst derivative method is developed as a focal quality indicator that measures image sharpness in the frequency spatial domain so that the computational burden introduced by the numerous two-dimensional DFTs could be greatly reduced. The topology of the burst derivative appears to be quite different from that of the entropy. The burst derivative has a quasi-periodic surface, so there are many local minima in addition to the desired global minimum [4].

A detailed analysis of the burst derivative presented next shows that the value of burst derivative changes as a function of motion parameter estimates. The definition given in (3.22) is used as a generalized burst derivative. For a point target whose signature has been partially motion compensated, the burst derivative can be written as

$$B = \sum_{m=0}^{M-2} \sum_{n=0}^{N-1} 4 \sin^2\left[4\pi \frac{f_n}{c}\left(r_{m,n} - \tilde{r}_{m,n} + r_{m+1,n} - \tilde{r}_{m+1,n}\right)\right] \tag{3.24}$$

where $r_{m,n} - \tilde{r}_{m,n}$ is defined as

$$r_{m,n} - \tilde{r}_{m,n} = r_0 + \Delta v t_{m,n} + \frac{1}{2}\Delta a t_{m,n}^2 \tag{3.25}$$

where $\Delta v$ is the velocity estimation error and $\Delta a$ is the acceleration estimation error.

### 3.4.1 Velocity Estimation Error

For a target with constant velocity, (3.24) becomes

$$G = \sum_{m=0}^{M-2} \sum_{n=0}^{N-1} 4 \sin^2\left(2\pi \frac{f_n}{c} \Delta v \Delta t N\right) \tag{3.26}$$

Notice that the right side is a summation of sinusoids, where each is a function of a different frequency $f_n$. In general, the double summation will not be zero for any value of $\Delta v$ except for $\Delta v = 0$. An exception to this occurs when the frequency step is an integer multiple of $f_0$. Consequently, the burst derivative has a global minimum and multiple local minima as shown in Figure 3.2. The figure shows that the presence of a global minimum at the actual target

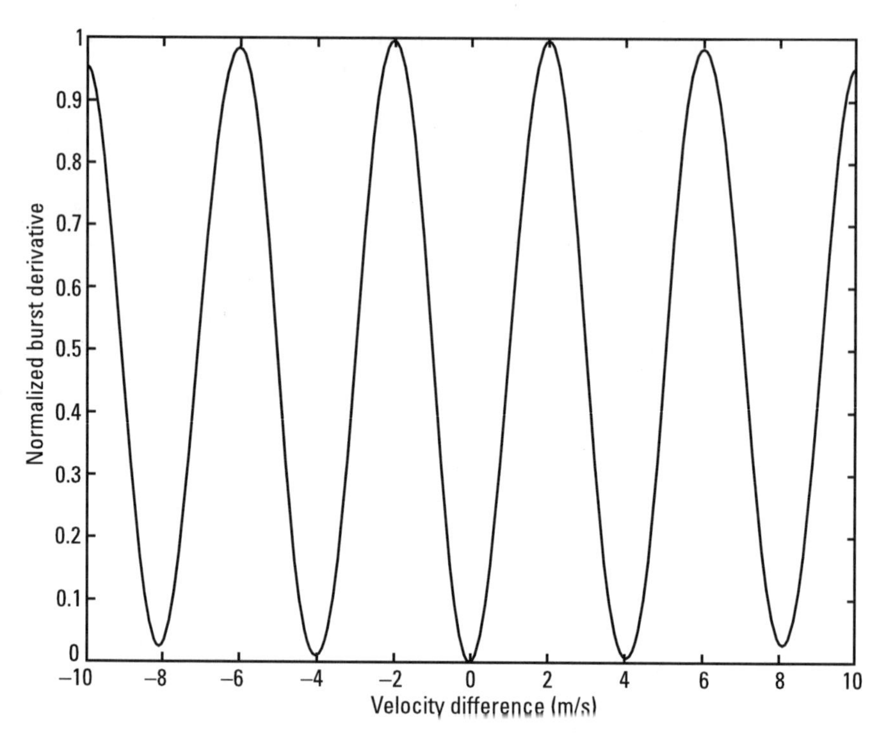


**Figure 3.2** Burst derivative as a function of velocity estimation error.

velocity and that the burst derivative $B$ is, in effect, an unbiased estimator of $\Delta v$ in the absence of noise.

The existence of many local minima in (3.26) precludes an optimum search of the target velocity; therefore, a priori knowledge regarding the velocity search space is required. For instance, an initial estimate may be obtained by using the entropy method, which is useful when several frames of the same target are to be obtained. The first frame is processed using the entropy approach, while the subsequent frames are processed by the burst derivative method. The velocity estimate of the first frame is used as an initial guess for the burst derivative method in subsequent frames.

For a small fractional bandwidth, (3.26) is approximated as

$$B = \sum_{m=0}^{M-2} \sum_{n=0}^{N-1} 4 \sin^2(2\pi \Delta v \Delta t N/\lambda) = 4(M-1)N \sin^2(2\pi \Delta v \Delta t N/\lambda) \tag{3.27}$$

where $\Delta v$ is the center wavelength of the radar. In this case, the maximum value of $B$ occurs when the argument of the sine is $\pi/2$. Furthermore, the nearest local minima occur at multiples of

$$\Delta v = \frac{\lambda}{2\Delta tN} \tag{3.28}$$

Thus, the search space must be confined to the region

$$-\frac{\lambda}{4\Delta tN} \leq \Delta v \leq \frac{\lambda}{4\Delta tN} \tag{3.29}$$

An accurate target velocity estimate is attainable only if the initial estimate of the velocity is within these limits. Therefore, the velocity estimation is severely limited by the initial estimate assumed in the search algorithm.

### 3.4.2 Acceleration Estimation Error

For the case of a uniformly accelerating point target, the dependence of the burst derivative on the acceleration estimation error is given by

$$B = \sum_{m=0}^{M-2} \sum_{n=0}^{N-1} 4 \sin^2\left[2\pi \frac{f_n}{c} \Delta a(\Delta tN)^2\left(1 + 2\frac{n + mN}{N}\right)\right] \tag{3.30}$$

In this case the initial target velocity is assumed to be zero, which is highly unlikely in practice. However, the zero velocity assumption simplifies the mathematical derivation. Equation (3.30) shows that the burst derivative becomes zero when the acceleration estimate is equal to the actual acceleration value ($\Delta a = 0$). This expression for the burst derivative as a function of acceleration estimate cannot be written in simpler form because the argument of the sine in (3.30) involves the indices $m$ and $n$. However, the positions of the local minima appear to be a quasi-periodic in terms of the acceleration error as shown in Figure 3.3.

## 3.5 Dependence of Burst Derivative on Radar Parameters

Figures 3.4 and 3.5 show the dependence of the burst derivative on radar parameters. In this instance, the limits of the velocity search space are $-6$ m/s and

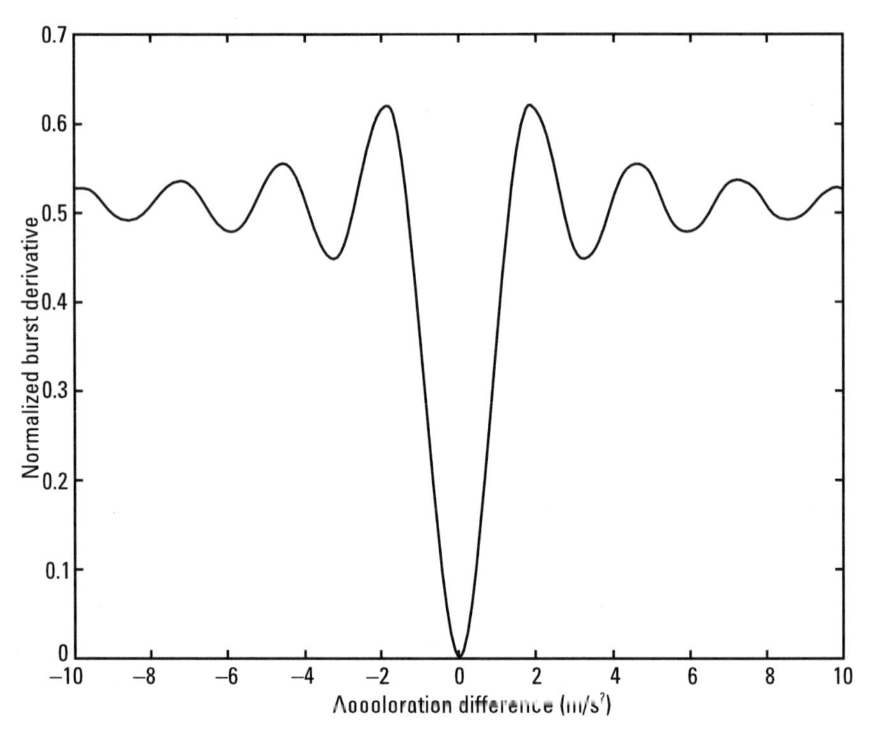


**Figure 3.3** Burst derivative as a function of acceleration estimation error.

6 m/s, respectively, to include several minima. Sufficient points are calculated to provide a velocity resolution of 0.05 m/s in all the plots.

Figure 3.4 shows the behavior of the burst derivative in terms of velocity and PRI. The values of PRI considered are 185 μs (solid line), 139 μs (dashed line), 92.6 μs (dashed-dotted line), and 46.3 μs (dotted line). The corresponding PRF values are 5.4 KHz, 10.5 KHz, 16.2 KHz, and 21.6 KHz. The most notable feature in Figure 3.4 is that the number of local minima decreases when the PRI decreases.

Figure 3.5 shows the behavior of the burst derivative when the number of pulses per burst and the frequency step are changed simultaneously to maintain a constant frequency sweep. The number of pulses per burst is reduced from 64 (solid line) to 48 (dashed line), then to 32 (dashed-dotted line), and finally to 16 (dotted line), thus increasing the frequency step from 4.32 MHz to 5.76 MHz, 8.64 MHz, and 17.28 MHz, respectively. A remarkable feature in Figure 3.5 is that the number of pulses and number of local minima are proportionally related. Thus, a lower number of pulses results in fewer local minima. However, having fewer pulses per burst causes the range granularity of the tar-

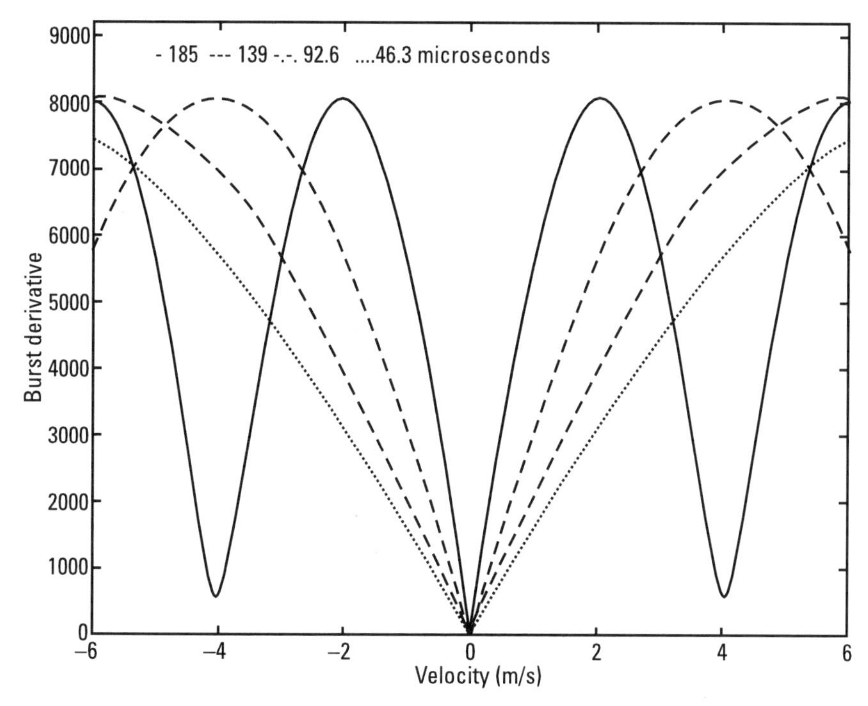


**Figure 3.4** Burst derivative as a function of PRI.

get's ISAR image to decrease and forces the unambiguous range window to lengthen.

## 3.6 Additive Noise

Under the assumption of additive Gaussian noise, the expected value of the burst derivative for a point target with translational motion is given by

$$B = 2N(M-1)\sigma^2 + \sum_{m=0}^{M-2}\sum_{n=0}^{N-1} 4\sin^2\left[4\pi\frac{f_n}{c}(r_{m,n} - \tilde{r}_{m,n} + r_{m+1,n} - \tilde{r}_{m+1,n}\right] \tag{3.31}$$

where

$$E(B) = 2N(M-1)\sigma^2 + B_0 \tag{3.32}$$

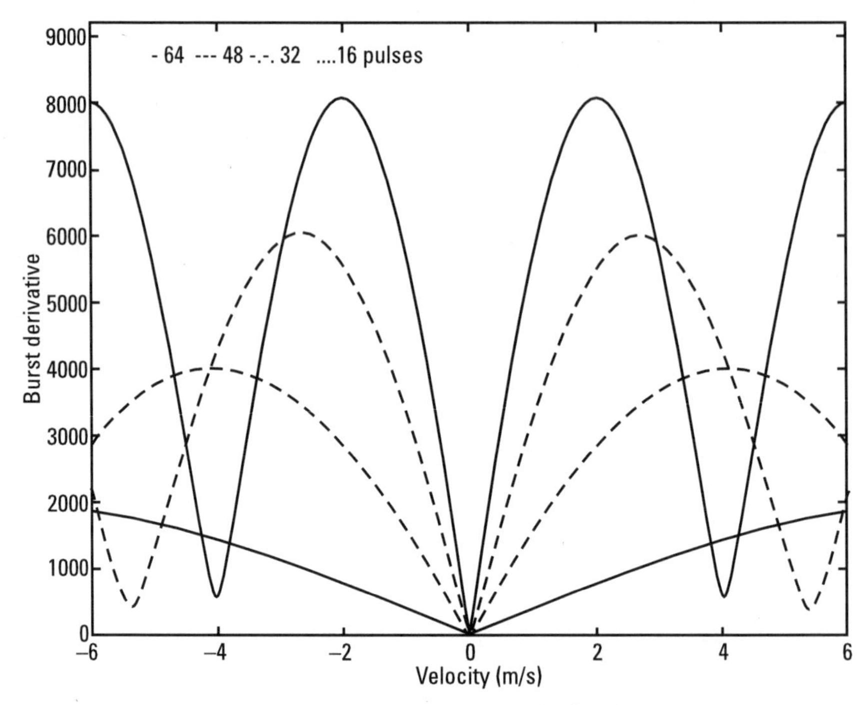


**Figure 3.5** Burst derivative as a function of the number of pulses in a burst.

Equation (3.32) shows that the expected value of the burst derivative in the presence of Gaussian noise is equal to the noise-free burst derivative $B_0$ plus a term dependent upon the variance of the additive noise. Thus, the expected value of the burst derivative is biased in the presence of noise.

## 3.7 Summary

This chapter demonstrated the connection between the burst derivative (a heuristic concept) and the Fisher information (a formal information concept) as applied to motion compensation. It showed that the burst derivative is strongly dependent on radar parameters such as PRI and number of pulses in a burst. The benefit of the burst derivative method is that it can be applied directly to the target reflectivity in the frequency domain, thus requiring significantly fewer computations. However, motion parameter estimation via the burst derivative method must deal with local minima in the search space and high SNR scenarios. In practice, this method can be used only within a narrow restricted search space. Thus, the disadvantages of the burst derivative method can be summarized as follows:

- Many local minima exist in the search space.
- The search space must be confined.
- In the presence of noise, the minimum value of $B$ will not be zero.

## References

[1] Bocker, R. P., and S. A. Jones, "ISAR Motion Compensation Using the Burst Derivative Measure as a Focal Quality Indicator," *Internat. J. of Imaging and Technology*, Vol. 4, 1992, pp. 285–297.

[2] Bar-Shalom, Y., and X. Li, *Estimation and Tracking: Principles, Techniques, and Software*, Norwood, MA: Artech House, 1993.

[3] Frieden, B. R., "Fisher Information, Disorder, and the Equilibrium Distributions of Physics," *Phys. Rev. A.*, 41(8), 1990, 4265–4276.

[4] Flores, B. C., S. Tariq, and J. S. Son, "Image Focus Quality Indicators for Efficient Inverse Synthetic Aperture Radar Phase Correction," *Proc. SPIE Algorithms for Synthetic Aperture Radar Imagery III*, Vol. 2757, 1996, pp. 2–13.

# 4

# Phase Difference Method

The burst derivative approach discussed in Chapter 3 is effective only when the search space is centered about its absolute minimum. If the search space is not confined because of lack of prior knowledge about target motion, then an exhaustive or complex optimization search method must be employed to find the location of minimal focal quality indicator. All those search methods are computationally expensive, resulting in slow motion compensation. Sometimes, it may be impossible to locate one global minimum location of many competing local minimal locations when SNR is low in echo signal. If a local minimum is not present in search space, then a simple search method can find the location of global minimum, and the motion parameter estimation process can be accelerated.

This chapter presents a new method for estimating the translational motion parameters of a target without restricting the search space. The basis for this method is a phase difference indicator [1] that converges to an absolute minimum when the phase is compensated with the actual values of the motion parameters. The method exploits the phase of the target's frequency response and does not require two-dimensional Fourier processing, thus making it superior to the entropy and burst derivative methods.

## 4.1 Phase of Target Response

One of the main objectives of motion compensation is to obtain an estimate of the target echo transfer function, $H(m, n)$, from the measured target response, $U(m, n)$. The process can be described analytically by considering a single point target with radial motion relative to the radar. In this case, $H(m, n)$ becomes

$$H(m, n) = h(x', y') \exp[j2\pi(p_{m,n}x' + q_{m,n}y')] \tag{4.1}$$

If the point scatterer is assumed to be the centroid of the target moving along the LOS of the radar, then $H(m, n)$ is a constant with respect to $m$ and $n$. Consequently, the target response is simply given by

$$U(m, n) = \exp\left[j4\pi \frac{f_n}{c}(r_0 + r_{m,n})\right] \tag{4.2}$$

Substituting (2.5) into (4.2), $U(m, n)$ becomes

$$U(m, n) = \exp\left[j\frac{4\pi}{c}(f_0 + n\Delta f)(r_0 + v_0 t_{m,n} + a_0 t_{m,n}^2/2\right] \tag{4.3}$$

Equation (4.3) shows that motion information is implicit in the phase of the reflection data collected by the radar. Thus, it is logical to consider phase processing directly in order to estimate motion parameters. Note that the phase term in (4.3) may be greater than $2\pi$. However, the detected phase is wrapped around between $-\pi$ and $\pi$, that is, the detected phase is ambiguous. Using the phase of the first pulse as a reference, the phase of the following pulses becomes unambiguous and can be unwrapped easily. However, that requires that the phase difference between consecutive pulses in a burst must not exceed $\pm\pi$ radians, which is same as the echo signal has to be sampled to its phase can be unwrapped unambiguously.

The reason that the burst derivative approach shows many local minima in its search space is that the method uses the wrapped phases to find the focal quality indicator. For the lapse that the phase does not exceed from the actual minimum location, there is not any local minimum. However, when it exceeds $\pm\pi$, the focal indicator value decreases, making local minimum. That local minimum repeats each time the phase difference exceeds from a local minimum, making multiple local minimum locations. The problem can be corrected by unwrapping the phase; thus a unique global minimum value without local minimum is expected using phase difference concepts.

## 4.2 Phase Difference Concepts

Suppose rough estimates of the initial velocity and acceleration of a target are available and those estimates are denoted as v and a. The motion compensation of the phase using those values leads to a phase term

$$\phi(m, n) = 4\pi \frac{f_n}{c}(r_0 + \Delta r_{m,n}) \tag{4.4}$$

where $\Delta r_{m,n}$ as defined in (2.20) is a range walk term. The compensated phase of the first pulse for the $m$th burst is

$$\phi(m, 0) = 4\pi \frac{f_0}{c}(r_0 + \Delta r_{m,0}) \tag{4.5}$$

Now, define the unambiguous, compensated phase for a pulse as

$$\phi'(m, n) = \phi(m, n) - \phi(m, 0) \tag{4.6}$$

and substituting (4.4) and (4.5) into (4.6) yields

$$\phi'(m, n) = \frac{4\pi}{c}[f_0(\Delta r_{m,n} - \Delta r_{m,0}) + n\Delta f(r_0 + \Delta r_{m,n})] \tag{4.7}$$

The phase term given in (4.7) is unwrapped with respect to the phase of the first pulse. However, note that the unknown range $r_0$ is still present in the equation.

Ideally, the phase history of each burst for a stationary target should be a linear function of pulse index and have the same slope for all bursts. Target motion introduces time-dependent nonlinearities. For proper phase compensation, the task is to recover the same linear phase history for each burst so that the slope of the phase history for each burst is the same. At that point, the initial phase for each burst is irrelevant. Therefore, the initial phase of all bursts can be shifted to an arbitrary value without any loss of information. That means the phase plot for each pulse will start at the same value, leaving only phase differences due to slope variation. If the phase is compensated using the actual values of the motion parameters, all the phase lines would be aligned with one another. Thus, an indicator of motion compensation can be obtained by taking a difference for a given pulse index as shown next:

$$P_i = \phi'(m, 0) - \phi'(m + 1, n) \tag{4.8}$$

or

$$P_i = \frac{4\pi}{c}[f_0(\Delta r_{m,n} - \Delta r_{m+1,n} - \Delta r_{m,0} + \Delta r_{m+1,0}) + n\Delta f(\Delta r_{m,n} + \Delta r_{m+1,n})] \tag{4.9}$$

Note that the unknown initial range has been eliminated by means of the difference operation. Moreover, when this difference reduces to zero for all pulse and burst indices, perfect motion compensation is achieved.

The above arguments lead to the definition of a phase difference indicator, $P$, given by the sum of the differences measured over all pulse and burst indices:

$$P = \sum_{m=0}^{M-2} \sum_{n=0}^{N-1} |P_i| = \sum_{m=0}^{M-2} \sum_{n=0}^{N-1} |\phi'(m, n) - \phi'(m+1, n)| \tag{4.10}$$

This indicator can be viewed as a surface over a multidimensional motion parameter space. Therefore, the process of motion compensation is that of finding the location of the global minimum.

## 4.3 Phase Difference Algorithm

The algorithm for motion compensation via the phase difference indicator $P$ can be summarized as follows:

1. Select estimates for the initial velocity and acceleration.
2. Compensate the phase of the frequency response using the estimates in step 1.
3. Unwrap the compensated phase in step 2. Note that the unwrapping process can be done by adding $2\pi k$ radians every time a phase discontinuity is found.
4. Shift the unwrapped phase for each burst so that the initial phase of each burst is always zero.
5. Calculate the phase difference indicator using (4.10).
6. Obtain improved estimates of the motion parameters and repeat steps 2 through 5 until the phase difference indicator $P$ is reduced to a minimum.
7. Generate a focused ISAR image.

## 4.4 Phase Difference as a Function of Motion Estimates

This section and Section 4.5 present results of analysis for two particular cases of motion compensation, each addressing the behavior of the phase difference indicator in one of the two possible dimensions of the search space. The analysis shows that the phase difference indicator has a single minimum in each dimension at a point where the compensating parameter estimate approaches the actual values of the target's initial velocity and acceleration.

### 4.4.1 Velocity Estimation Error

Assume that a point target moves on LOS at a constant radial velocity $v_0$. By compensating the phase with a velocity estimate $v$, the range walk is reduced to $\Delta r_{m,n}$. In this case, $\Delta r_{m,n}$ can be written as

$$\Delta r_{m,n} = \Delta v t_{m,n} + \Delta v(n + mN)\Delta t \tag{4.11}$$

where $\Delta v = v_0 - v$. Using (4.9), (4.10), and (4.11), the expression for the phase difference indicator is reduced to

$$P = \sum_{m=0}^{M-2} \sum_{n=0}^{N-1} \frac{4\pi}{c} nN\Delta f \Delta t |\Delta v| \tag{4.12}$$

which can be expressed in a closed form as

$$P = \frac{2\pi}{c}(M-1)(N-1)N^2 \Delta f \Delta t |\Delta v| \tag{4.13}$$

Equation (4.13) shows that the phase difference indicator is directly proportional to the error in the velocity estimate. Notice that $P$ equals zero when the error is reduced to zero. Equation (4.13) is shown in Figure 4.1 as a function of velocity estimation error.

Unlike the burst derivative approach discussed in Chapter 3, the phase difference indicator does not show any local minimum as long as the Nyquist condition is satisfied when the echo signal is sampled.

### 4.4.2 Acceleration Estimation Error

Assume that a point target moves on the LOS with a constant acceleration and let $\Delta v = 0$. That condition is unlikely to happen in practice because the

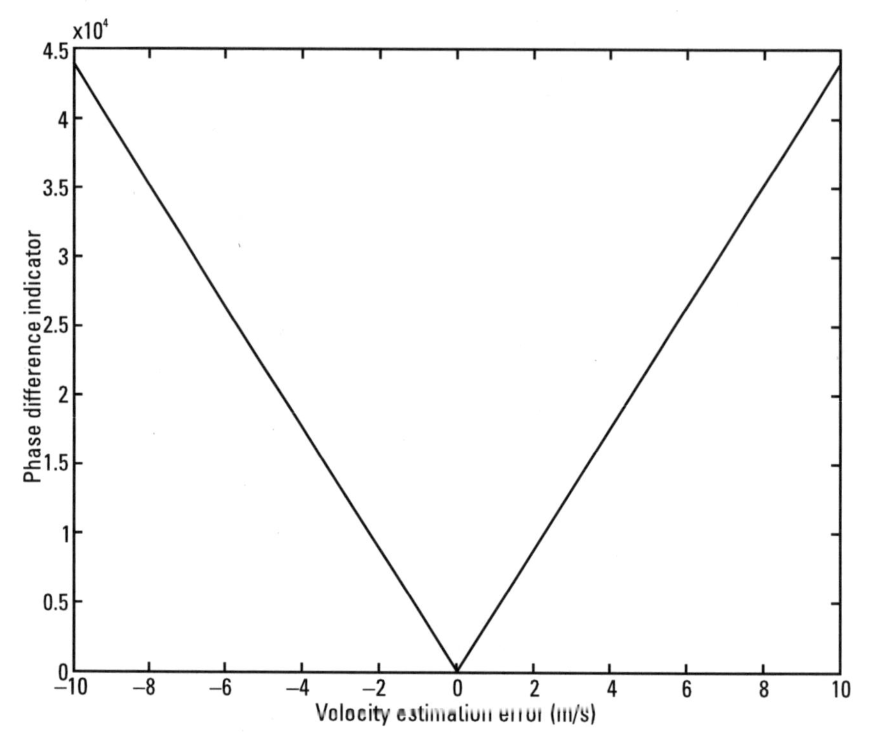


**Figure 4.1** Phase difference indicator as a function of velocity estimation error.

acceleration estimation error introduces velocity estimation error and vice versa. However, assumption of that condition gives insight about the acceleration effect on the phase difference indicator. The condition simplifies the range walk equation as:

$$\Delta r_{m,n} = \frac{1}{2}\Delta a t_{m,n} \tag{4.14}$$

where $\Delta a = a_0 - a$. Using (4.9), (4.10) and (4.14), we can express the phase difference indicator as

$$P = \sum_{m=0}^{M-2} \sum_{n=0}^{N-1} \frac{2\pi}{c} \Delta t^2 |\Delta a| (2 f_0 nN + 2n^2 N \Delta f + 2mnN^2 \Delta f + nN^2 \Delta f) \tag{4.15}$$

which yields

$$P = \frac{2\pi}{c}\Delta t^2 |\Delta a| (M-1)(N-1)N^2$$

$$\left[f_0 + \frac{1}{3}(2N-1)\Delta f + \frac{1}{2}(M-2)N\Delta f + \frac{1}{2}N\Delta f\right] \quad (4.16)$$

Here, the phase difference indicator is a linear function of the acceleration estimate error. The similarity between (4.16) and (4.13) is clear. Thus, the argument for a unique solution is applicable for this case as well. Equation (4.16) is shown in Figure 4.2 as a function of the acceleration estimation error.

## 4.5 Results

The proposed phase difference method is designed to overcome the shortcomings of the entropy and burst derivative approaches. In this approach, Fourier

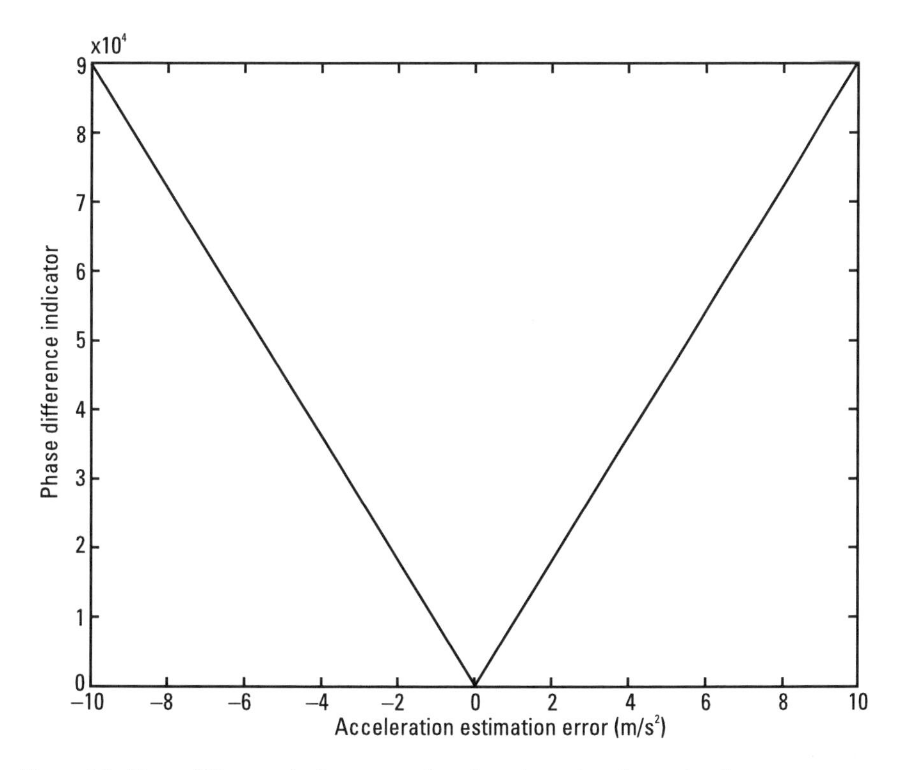


**Figure 4.2** Phase difference indicator as a function of acceleration estimation error.

processing is bypassed altogether except when the final image is obtained. Furthermore, the phase difference indicator shows a unique minimum in high SNR. We applied the phase difference method to an ISAR signature corresponding to an ideal point scatterer and to an actual DC-10 airplane signature. In both cases, the phase difference measurement converged to a unique minimum focal quality indicator values. Figure 4.3 shows the image quality measurement using the phase difference for a point scatterer. The actual target velocity and acceleration values were $-31.5$ m/s and 3 m/s$^2$, respectively, which correspond the center of the plot. Figure 4.3 shows that this method can accurately estimate the target motion parameters for a noise free point target.

When the method was applied to an actual Boeing 727 ISAR signature, the phase difference indicator value did not provide reliable motion parameters. One of the factors that affect this method is the discontinuity in the unwrapped phase of actual data due to noise. Especially the reflection phase changes rapidly, even for a small aspect change in the target [2]. If the noise is added to the radar echo signal, the phase changes even faster, making the phase unwrapping process ambiguous. Thus, the phase difference method is sensitive to noise, and its convergence depends on the behavior of the ISAR signature.

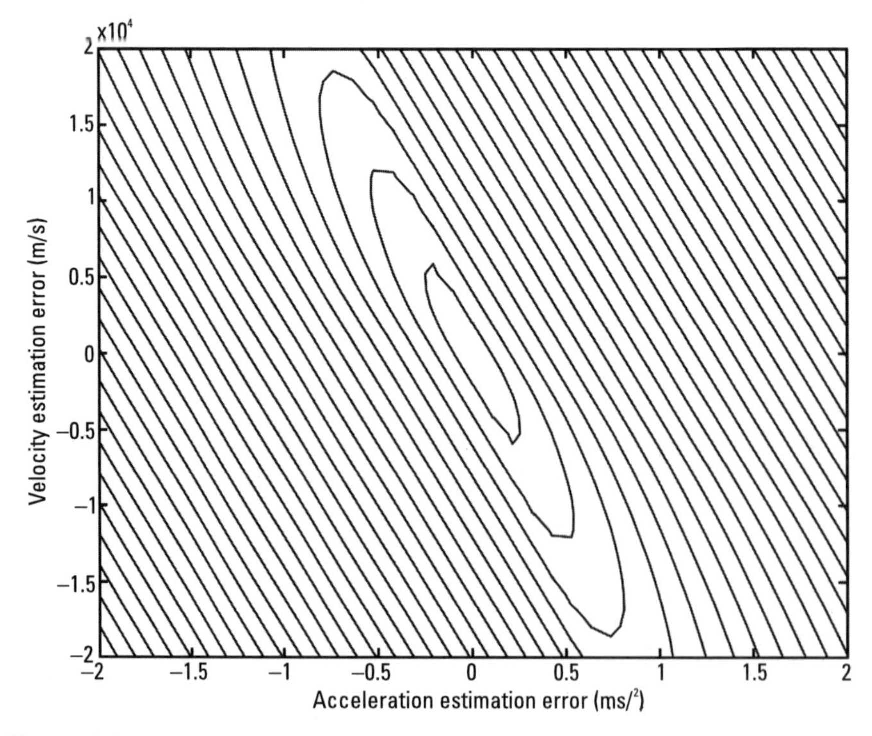


**Figure 4.3** Image quality measurement using phase difference.

## 4.6 Summary

This chapter demonstrated the potential use of the phase difference method for motion compensation. It showed that the target range information contained in the phase can be extracted through phase slope measurements without two-dimensional DFT processing. The phase slope is determined by the unwrapped phase difference between two adjacent bursts. A new focal quality indicator was developed based on the summation of the derivative of those phase differences. This focal quality indicator reaches a minimum when the phase difference remains constant. It was shown that the motion parameter estimates are exact for infinite SNR point scatterers. However, the focal quality indicator showed some local minima when the method was tested with low SNR signatures or actual targets, thus showing that the method is sensitive to noise. Thus, the phase difference method has two key disadvantages:

- The method is highly sensitive to noise.
- The convergence depends on the behavior of the ISAR signature's phase.

## References

[1] Son, J. S., S. Tariq, and B. C. Flores, "Phase Difference Method for Target Motion Compensation of Stepped-Frequency ISAR Signature," *Proc. SPIE Radar Processing, Technology, and Applications,* Vol. 2845, 1996, pp. 163–174.

[2] Wehner, D. R., *High-Resolution Radar,* Norwood, MA: Artech House, 1995.

# 5

# Least Squares Motion Parameter Estimation

The phase difference method discussed in Chapter 4 is based on a focal quality indicator that can be interpreted as the summation of increments of the unwrapped phases. This chapter deals with a more noise resistant motion estimation approach based on least squares estimation that utilizes a focal quality indicator possessing a unique global minimum with no local minima for a wide SNR dynamic range. This approach measures the relative range of directly from the phase of the echo transfer function for accuracy effectiveness, noise resistance. Also this chapter provides the theoretical foundation of the application of the least squares method to the wideband radar motion estimation, and serves as an introduction to the more generalized least squares method which is discussed in a latter chapter.

## 5.1 General Least Squares Problem

A typical wideband radar signature record is an over-determined sample set that can be approximated by a linear equation matrix. A unique solution linear equation matrix does not exist. A closed form solution can be found by combining parameters for which the sum of the squares of equation errors is a minimum, known as least squares estimation.

The fundamental least squares problem involves an over-determined set of linear equations expressed as

$$\mathbf{Ax} = \mathbf{b} \tag{5.1}$$

where the $\mathbf{A}$ is $m \times n$ and where $m > n$. Again, since over-determined equations are inconsistent due to measurement errors, it is not possible to find a unique solution for $\mathbf{x}$. An approximate solution, denoted by $\hat{\mathbf{x}}$, can be obtained by requiring that the sum of the squares of the errors between $\mathbf{b}$ and $\mathbf{A}\hat{\mathbf{x}}$, viz.,

$$J(\hat{\mathbf{x}}) = (\mathbf{b} - \mathbf{A}\hat{\mathbf{x}})^t(\mathbf{b} - \mathbf{A}\hat{\mathbf{x}}) = \mathbf{e}^t\mathbf{e} \tag{5.2}$$

be minimized. This implies that the estimate $\hat{\mathbf{x}}$ is given by

$$\mathbf{Ax} + \mathbf{e} = \mathbf{b} \tag{5.3}$$

where the sum of the squares of the elements of the error vector $\mathbf{e}$ is a minimum.

To find the minimum of $J(\hat{\mathbf{x}})$, its partial derivatives with respect to each of the elements of $\mathbf{A}\hat{\mathbf{x}}$ must be zero. According to [1, 2], the derivative of a scalar with respect to a vector is given by

$$\frac{\partial}{\partial \mathbf{x}}(\mathbf{x}^t\mathbf{y}) = \frac{\partial}{\partial \mathbf{x}}(\mathbf{y}^t\mathbf{x}) = \mathbf{y}^t \tag{5.4}$$

or

$$\frac{\partial}{\partial \mathbf{x}}(\mathbf{x}^t\mathbf{A}\mathbf{x}) = \mathbf{x}^t\mathbf{A}^t + \mathbf{x}^t\mathbf{A} \tag{5.5}$$

where the superscript $t$ denotes the transpose of a matrix.

Applying (5.4) and (5.5) to the partial derivative of (5.2) yields

$$\left(\frac{\partial J}{\partial \mathbf{x}}\right)^t = \mathbf{A}^t(\mathbf{b} - \mathbf{A}\hat{\mathbf{x}}) = 0 \tag{5.6}$$

From (5.4), it follows that

$$\mathbf{A}^t\mathbf{A}\hat{\mathbf{x}} = \mathbf{A}^t\mathbf{b} \tag{5.7}$$

so that the approximate solution vector can be written as

$$\hat{\mathbf{x}} = (\mathbf{A}^t\mathbf{A})^{-1}\mathbf{A}^t\mathbf{b} \tag{5.8}$$

If **A** is full rank, the least square estimate $\hat{\mathbf{x}}$ given by (5.8) is unique. In addition to giving a unique solution for a given data set, the least squares method is also in agreement with the maximum likelihood principle of statistics [3]. Thus the least squares estimate $\hat{\mathbf{x}}$ is interpreted as the most likely value of the vector **x**.

## 5.2 Fitting a Polynomial to a Set of Measurements

Polynomial fitting for target tracking is discussed in detail in [4]. As a brief introduction to the polynomial fitting, let's assume the position of a target can be modeled as a polynomial in time. That is,

$$\xi(t) = \sum_{j=0}^{n-1} x_j \frac{t^j}{j!} \tag{5.9}$$

where the parameters estimated are the polynomial coefficients $x_j$, $j = 0, 1, \ldots,$ $n - 1$. The least squares technique from the previous section will be used to estimate these parameters via a polynomial fitting of order $n$.

The measurements of the target position can be written as

$$b(i) = a(i)'x + e(i) \tag{5.10}$$

where

$$x(i)' = [x_0 \quad x_1 \ldots x_{n-1}]' \tag{5.11}$$

is the $n$-dimensional parameter vector to be estimated and the row vector is

$$a(i)' = \left[1 \quad t_i \ldots \frac{t_i^n}{n!}\right] \tag{5.12}$$

The measurement noise $e(i)$ is assumed to be a zero-mean Gaussian noise with variance $\sigma^2$, Then the measurement matrix is

$$\mathbf{A} = \begin{bmatrix} a(1)' \\ \vdots \\ a(m)' \end{bmatrix} \tag{5.13}$$

From (5.8), the estimate of the parameter vector $\mathbf{x}$ is

$$\hat{x}(m) = \left\{ \sum_{i=1}^{m} a(i)a(i)' \right\}^{-1} \sum_{i=1}^{m} a(i)b(i) \tag{5.14}$$

with the covariance matrix

$$C(m) = \sigma^2 \left\{ \sum_{i=1}^{m} a(i)a(i)' \right\}^{-1} \tag{5.15}$$

Note that the term to be inverted in (5.14) and in (5.15), which is $n \times n$ matrix, requires $m > n$ in order for the inverse to exist. Consequently many measurements must be greater than the number of the parameters are needed.

## 5.3 Application of Least Squares to Motion Compensation

Polynomial fitting can be incorporated into the computation of a focal quality indicator to reduce estimation errors caused by the finite order of the parametric motion model. Consider the phase equation given in (4.4). Introducing the quadratic time dependent range variable given in (2.20) into (4.4) yields

$$\phi(m, n) = 4\pi \frac{f_n}{c} \left( r_0 + \Delta v t_{m,n} + \frac{1}{2} \Delta a t_{m,n}^2 \right) \tag{5.16}$$

Notice that the phase term due to $r_0$ is present in every burst. This phase term can be easily eliminated by subtracting the phase of the first burst from the phase of subsequent bursts. Since the sampling time for the first burst is

$$t_{0,n} = n \Delta t \tag{5.17}$$

The phase function of the first burst can be written as

$$\phi(0, n) = 4\pi \frac{f_n}{c}\left(r_0 + n\Delta v\Delta t + \frac{1}{2}n^2\Delta a\Delta t^2\right) \tag{5.18}$$

the phase difference equation referenced to the first burst can be written as

$$\phi(m, n) - \phi(0, n) = 4\pi \frac{f_n}{c}\left(\Delta v(t_{m,n} - n\Delta t) + \frac{1}{2}\Delta a(\Delta t_{m,n}^2 - n^2\Delta t^2)\right) \tag{5.19}$$

or

$$\phi(m, n) - \phi(0, n) = 4\pi \frac{f_n}{c}\left(mN\Delta v\Delta t + \frac{1}{2}\Delta a(2mnN + m^2N^2)\Delta t^2\right) \tag{5.20}$$

Introducing (2.7) into (5.20) yields

$$\phi(m, n) - \phi(0, n) = \frac{4\pi}{c}[mNf_0\Delta v\Delta t + mnN\Delta f\Delta v\Delta t$$

$$+ \Delta a(2mnNf_0 + m^2N^2f_0 + m^2N^2f_0 + 2n^2nM\Delta f + nm^2N^2\Delta f)\Delta t] \tag{5.21}$$

When an ideal point target is stationary, the phase history of each burst is a linear function of the pulse index and the phase slope associated with all the bursts is constant. Since target motion introduces distortion in the phase, the phase slope is not the same from burst to burst. Recall that the phase slope is determined by the phase difference between two adjacent pulses. Thus, the phase slope is a function of $n$. Therefore any term which is not a function of $n$ in (5.21) can be ignored. Let us denote the phase slope $\phi(m, n)$ as

$$\phi(m, n) = \frac{4\pi}{c}$$

$$\times [mnN\Delta f\Delta v\Delta t + {}^1/_2\Delta a(2mnNf_0 + 2n^2mN\Delta f + nm^2N^2\Delta f)\Delta t] \tag{5.22}$$

The phase of the last pulse in a burst is

$$\phi(m, N) = \frac{4\pi}{c}$$

$$\times [mN^2\Delta f \Delta v \Delta t + {}^1\!/_2 \Delta a(2mN^2 f_0 + 2mN^3\Delta f + m^2N^3\Delta f)\Delta t] \tag{5.23}$$

Notice that three terms are multiplied by $\Delta a$. Considering that

$$m^2N^3\Delta f \gg 2mN^3\Delta f + 2mN^2 f_0 \tag{5.24}$$

for $m > 3$, equation (5.23) can be simplified to

$$\phi(m, N) = \frac{4\pi}{c}\left(mN^2\Delta f \Delta v \Delta t + \frac{1}{2}\Delta a m^2 N^3 \Delta f \Delta t^2\right) \tag{5.25}$$

Also, notice that the time duration of a burst is $\Delta\tau = N\Delta t$, and that the time associated with the $m$th burst is $t_m = mN\Delta t$. Thus, (5.25) can be written as

$$\Phi(m) = (4\pi N\Delta f / c)\left(\Delta v t_m + \frac{1}{2}\Delta a \Delta t_m^2\right) \tag{5.26}$$

By setting $K = (4\pi/c)N\Delta f$, (5.26) becomes

$$\Phi(m) = K\left(\Delta v t_m + \frac{1}{2}\Delta a \Delta t_m^2\right) \tag{5.27}$$

Now the objective is to find the parameters $K\Delta v$ and $(K\Delta a)/2$ using a second order poynomial fitting scheme. Hence the adopted model is:

$$y = \alpha_1 + \alpha_2 t_m + \alpha_3 t_m^2 \tag{5.28}$$

Let $\alpha_2 = K\Delta v$ and $\alpha_3 = K\Delta a/2$ and define the matrix and vectors as follows:

$$\mathbf{A} = \begin{bmatrix} 1 & \Delta\tau & \Delta\tau^2 \\ 1 & 2\Delta\tau & (2\Delta\tau)^2 \\ 1 & 3\Delta\tau & (3\Delta\tau)^2 \\ \vdots & \vdots & \vdots \\ 1 & M\Delta\tau & (M\Delta\tau)^2 \end{bmatrix} \tag{5.29}$$

$$\mathbf{b} = \Phi \tag{5.30}$$

$$\hat{\mathbf{x}} = \alpha \tag{5.31}$$

By substituting (5.29), (5.30) and (5.31) into (5.8), a unique solution for $\alpha$ is obtained.

Now we define a new focal quality indicator

$$\gamma = [(\alpha_2)^2 + (\alpha_3)^2]^{1/2} = [(K\Delta v)^2 + (K\Delta a)^2]^{1/2} \tag{5.32}$$

which is to be minimized in the least squares sense. It is clear that the focal quality indicator in (5.32) will approach a minimum value when the actual values of velocity and acceleration are used to compensate the target's echo transfer function.

The algorithm was tested with simulated data, and it was verified that the focal quality indicator has a global minimum for the actual target motion parameters. Thus, the algorithm was applied to the Boeing 727 experimental data set. Figure 5.1 shows the results of an numerically exhaustive search using (5.32) based on the least squares approach. An exhaustive search was performed over the two dimensional space confined to $-14\ (m/s) \leq v \leq -8\ (m/s)$ and $0\ (m/s^2) \leq a \leq 5\ (m/s^2)$. The surface shows convexity toward the center of the search space. Here, the minimum focal quality indicator value has coordinates at $v = 11.3\ (m/s)$ and $a = 2.6\ (m/s^2)$.

However, when an image was formed after performing motion compensation using these motion parameter estimates, it was observed that the blurring was so severe that the aircraft could not be recognized. This, of course, is an indication that the location of the global minimum is significantly offset from its expected position in the velocity and acceleration plane. The main factor that contributes to this parameter estimation error is a series of phase jumps caused by noise interference. Another factor that contributes to the error is the amplitude scintillation observed in the actual data. Thus, the use of this motion compensation technique in its current formulation requires further improvement. This indicates that the unique global minimum does not guarantee the motion estimation parameters are actual target parameters. The focal quality

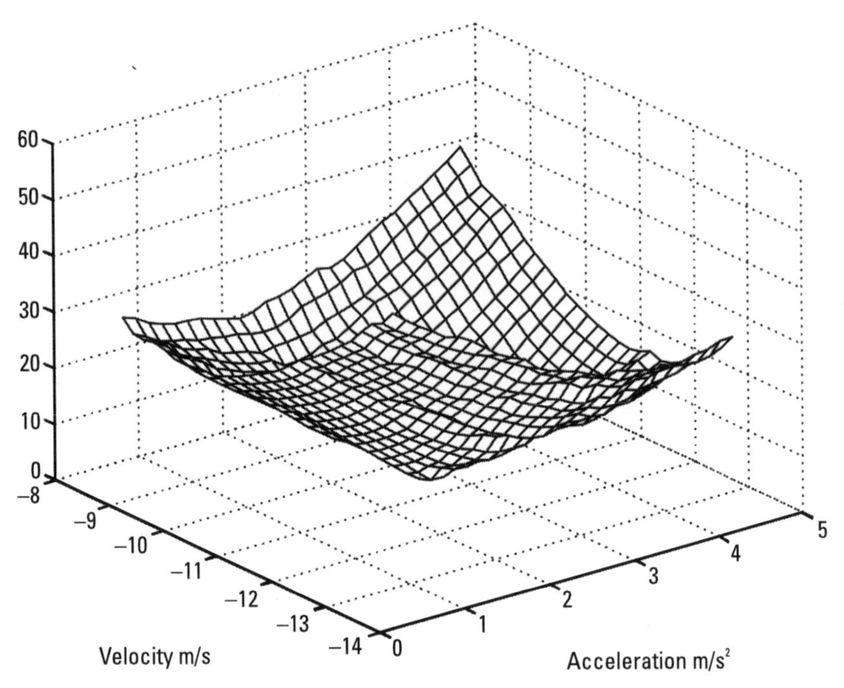


**Figure 5.1** Least squares based focal quality indicator for an actual Boeing 727 data record.

indicator must have not only a unique global minimum point in its search plane when the data is corrupted with noise but also must provide accurate motion parameter estimates.

## 5.4 Summary

A new focal quality indicator was defined by means of a least squares polynomial fitting of the phase slope function. Considering ideal point targets, the least squares method for motion estimation proposed in this chapter is more noise resistant than the phase difference method. In this current form, this focal quality indicator showed a minimum whose estimation coordinates match the actual values of the point target's initial velocity and acceleration. However, when the least squares method was applied to an actual target signature, phase slope discontinuities and amplitude scintillation led to inaccurate motion parameter estimates. The solution to this problem is the subject of discussion in the following chapter.

## References

[1] Hostetter, G. H., M. S. Santina and P. D'Carpio-Montalvo, *Analytical, numerical, and computational methods for science and engineering,* Englewood Cliffs, NJ: Prentice-Hall, 1991.

[2] Morris, J. L., *Computational Methods in Elementary Numerical Analysis,* John Wiley & Sons, 1983.

[3] Gerald, C. F., and P. O. Wheatley, *Applied Numerical Analysis,* Addison-Wesley, 1994.

[4] Bar-Shalom, Y., and X. Li, *Estimation and Tracking: Principles, Techniques, and Software,* Norwood, MA: Artech House, 1993.

# 6

# Complex Analysis of ISAR Signatures

In practice, the least squares approach for motion estimation and compensation is severely limited by its sensitivity to phase discontinuities and amplitude scintillation. This chapter presents a new phase slope measurement method that incorporates magnitude and phase measurements in computing a focal quality indicator whose minimum coordinates correspond to accurate target motion parameter estimates even for a signal which its magnitude fluctuate.

By its nature, the proposed complex analysis method is restricted to high-SNR cases. When noise filtering is required, a technique that exploits noise statistical properties to discern between noise and signal is applied. Here, we consider constant false alarm rate (CFAR) detection in range using a threshold determined by the Weibull distribution [1]. By using CFAR filtering, the noise level is suppressed sufficiently so that the phase slope is measured accurately. The measurement of phase slope is achieved using sections of the motion compensated target's frequency response in which its magnitude is high and its phase nonlinearity is small.

Conventional ISAR motion compensation techniques use the intensity of the target's range profile to measure the relative range of each scattering point. Those techniques are adequate if the target can be modeled as a point scatterer, justified only if each scatterer is small compared to the wavelength. In practice, the range profile of a man-made radar target is often too complicated to be adequately represented by this model. Figure 6.1 illustrates this situation for a Boeing 727.

Even if the model applies, there is still the problem of mutual interference among returns from different scatterers. Interference causes the reflectivity to fluctuate, as shown in Figure 6.2.

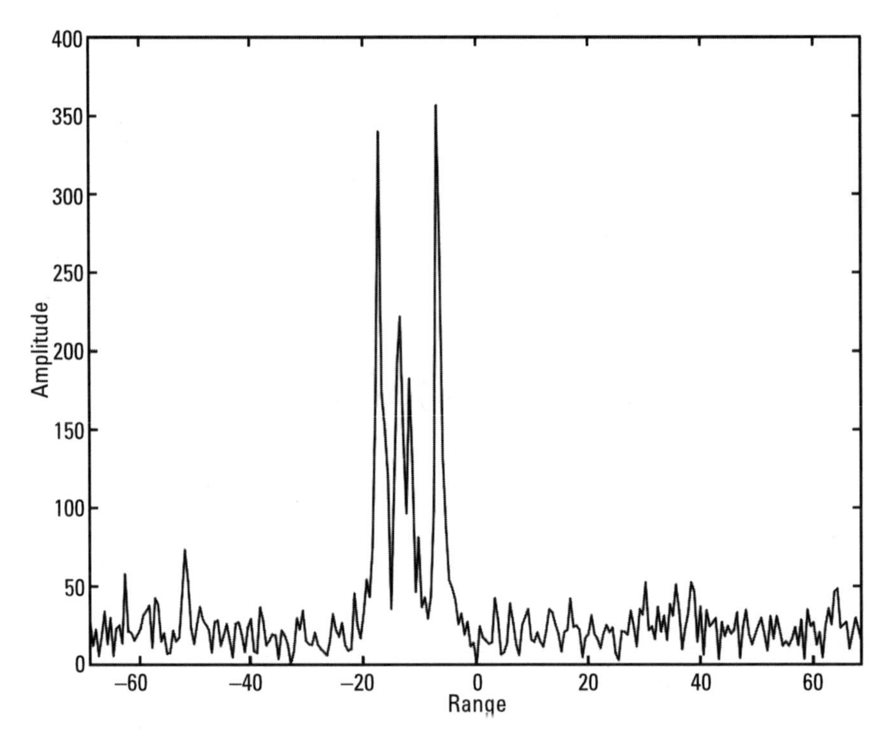


**Figure 6.1** Range profile of a Boeing 727.

## 6.1 Noise Suppression for Phase Slope Measurement

A complex analysis approach to measure the relative range of centroid of scatterers directly from the phase of the echo transfer function is proposed for accuracy and effectiveness. As mentioned in Section 4.2, an accurate measurement of the phase slope is tantamount to measuring instantaneous target range; thus the target motion estimation can be improved.

Figure 6.3 illustrates the unwrapped phase of the echo transfer function of the Boeing 727 signature. The figure shows linear segments and occasional phase fluctuations. Since the slope of the phase cannot oscillate for a target moving straight on the LOS during a short time interval, it must be assumed that the phase fluctuation occurs due to the noise in the echo signal. To reduce those fluctuations, the echo transfer function must be filtered and raise the SNR. This noise reduction process is done in the range domain using a finite impulse response (FIR) filter.

As an example, let us consider a Boeing 727 signature. We want to suppress noise outside the range window from −28m to 4m, which is one-fourth the original range window. The filtered profile is shown in Figure 6.4. The cor-

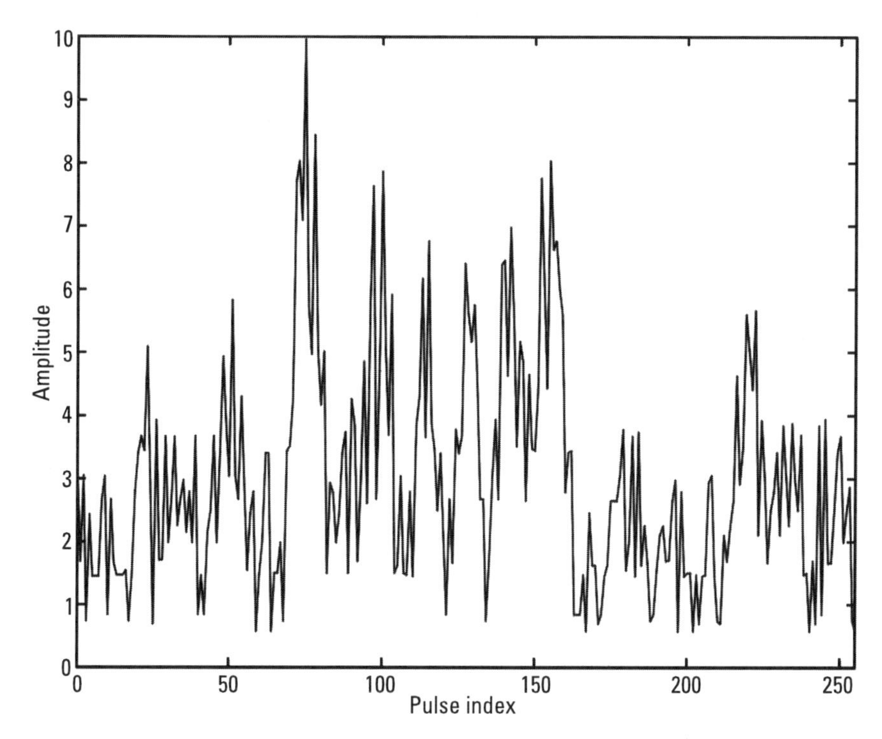


**Figure 6.2** Amplitude of Boeing 727 echo transfer function.

responding amplitude and phase of the echo transfer function, shown in Figure 6.5 and Figure 6.6, respectively, indicate that the fluctuations of the amplitude and phase responses are significantly reduced.

Consequently, an improved measurement of the target range can be performed. To estimate the range, we properly select intervals in the signal where the amplitude is high and the phase deviates little from a straight line. The straight phase line ensures that the measured phase slope is an accurate, instantaneous target range representation almost free of noise. In practice, the amplitude threshold for interval selection is no lower than 20 or 30% of the average amplitude, while phase deviations are no more than 0.1 or 0.2 cycle.

## 6.2 Noise Suppression Using CFAR Filtering

This section proposes CFAR filtering for noise suppression as an initial step toward motion estimation. The properties of the CFAR filter are to be determined in the range domain. Consequently, range profiles must be generated via the IDFT. Regions in the range profile dominated by noise are exploited to fit

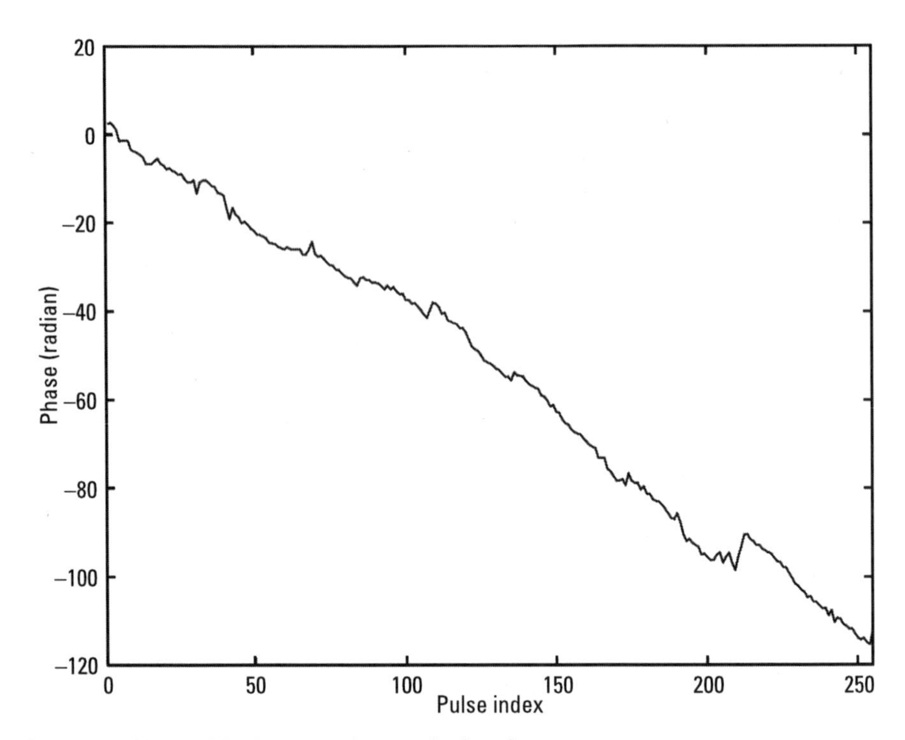


**Figure 6.3** Phase of Boeing 727 echo transfer function.

noise distribution to a particular PDF. For convenience, we chose the Weibull PDF to model the noise distribution. In contrast, regions dominated by scatterer signatures are used to determine the width of the filter.

### 6.2.1 Weibull Distribution

The Weibull distribution is a function of a scale and a shape parameter, defined as

$$f_X(x) = abx^{b-1}\exp(-ax^b) \quad \text{for} \quad x > 0 \tag{6.1}$$

where $a$ and $b$ are the parameters of the distribution. Because the Rayleigh PDF is a special case of the Weibull PDF, the output of an envelope detector from Gaussian distributed samples fits the model. That would occur when the shape parameter $b$ tends to be equal to 2. Another PDF described by the Weibull distribution is the exponential PDF. In that case, $b$ would tend to be equal to 1. It should be noted that the Weibull PDF has proved itself to be an effective model of sea and cloud radar clutter [1, 2].

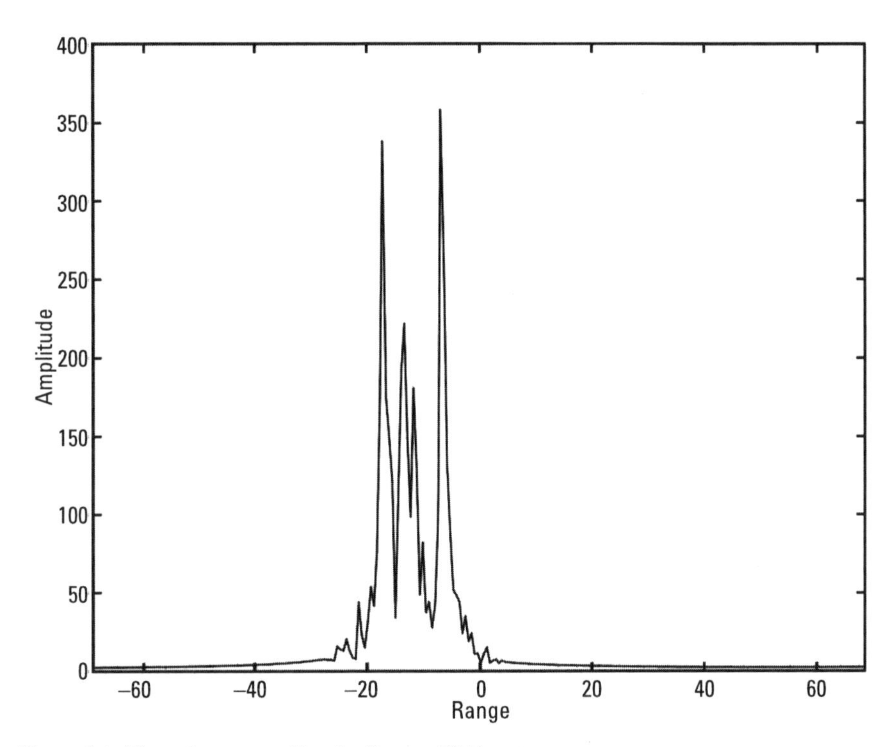


**Figure 6.4** Filtered range profile of a Boeing 727 burst.

### 6.2.2 Maximum Likelihood Estimation of Weibull Parameters

Given the noise samples $x_1, x_2, \ldots, x_n$, we seek to estimate the parameters $a$ and $b$ of the PDF in (6.1). The Maximum likelihood estimator (MLE) is considered first. Let the noise be a random sample from $f_X(x)$ as defined in (6.1). The likelihood function for $n$ samples of $x$ is expressed as

$$L = L(a, b) = \prod_{i=1}^{n} f_X(x_i; a, b) \tag{6.2}$$

Thus, the MLE of $a$, $b$ denoted as $\hat{a}$, $\hat{b}$ satisfies

$$L(\hat{a}, \hat{b}) \geq L(a, b) \quad \text{for all} \quad a, b \neq \hat{a}, \hat{b} \tag{6.3}$$

To find the estimates, the following system of equations needs to be solved:

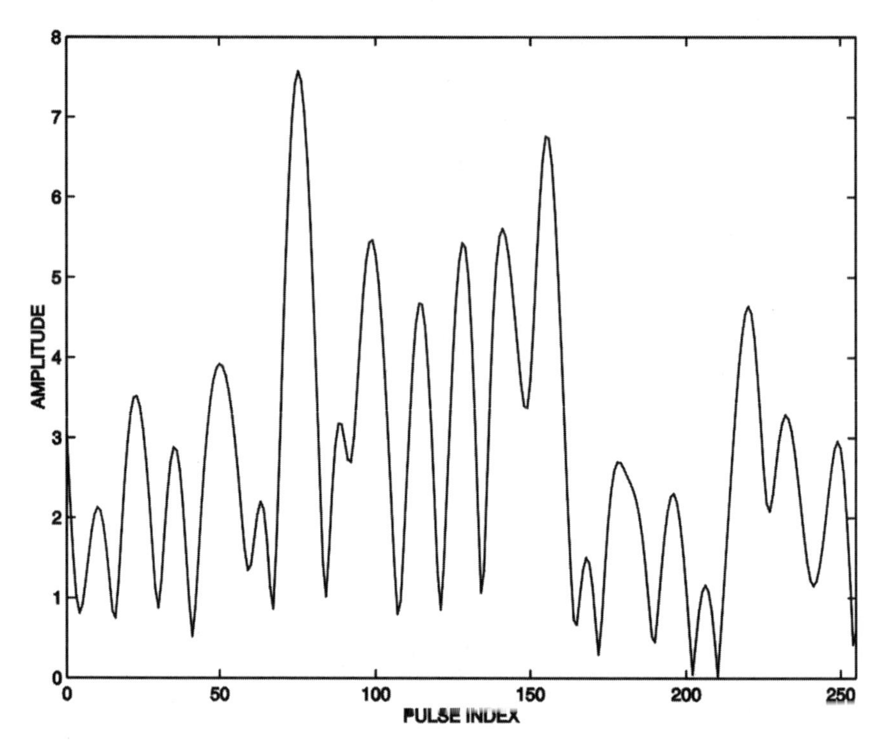


**Figure 6.5** Amplitude of filtered echo transfer function of Boeing 727 burst.

$$\frac{\partial L(a, b)}{\partial a} = 0 \quad \text{and} \quad \frac{\partial L(a, b)}{\partial b} = 0 \tag{6.4}$$

where $L(a, b)$ is defined by

$$L(a, b) = \prod_{i=1}^{n} abx_i^{b-1} e^{-ax_i^b} \tag{6.5}$$

Taking the logarithm

$$\ln L(a, b) = \sum_{i=1}^{n} \ln(abx_i^{b-1} e^{-ax_i^b}) \tag{6.6}$$

and the derivatives, we have the following expressions:

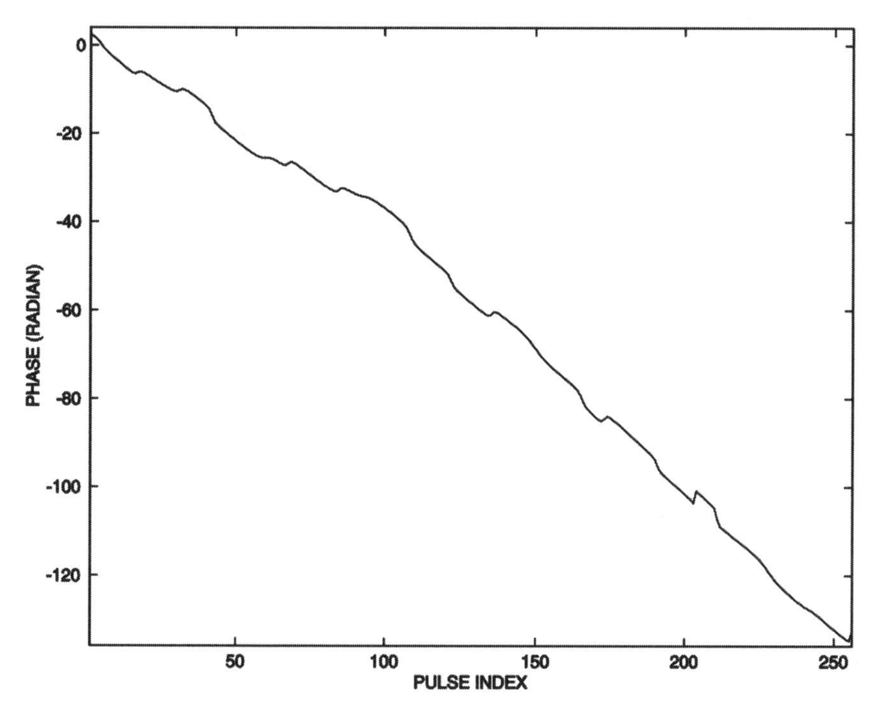


**Figure 6.6** Phase of filtered echo transfer function of Boeing 727 burst.

$$\frac{d[\ln L(a, b)]}{da} = \frac{n}{a} - \sum_{i=1}^{n} x_i^b = 0 \tag{6.7}$$

$$\frac{d[\ln L(a, b)]}{da} = \frac{n}{b} - a\sum_{i=1}^{n} x_i^b \ln(x_i) + \sum_{i=1}^{n} \ln(x_i) = 0 \tag{6.8}$$

No closed-form solution exists for solving $a$ and $b$; therefore, a numerical solution is needed that requires a considerable amount of computational time, depending on the amount of data available.

As indicated in Figure 6.7, this approach gives the best estimation results in terms of the mean square error (MSE). However, as shown in Figure 6.8, the approach employs a large number of floating point operations.

### 6.2.3 Weibull Parameter Estimation Via the Method of Moments

Another popular approach for parameter estimation is the method of moments, which consists of equating $k$ theoretical moments to the corresponding sample moments [3]:

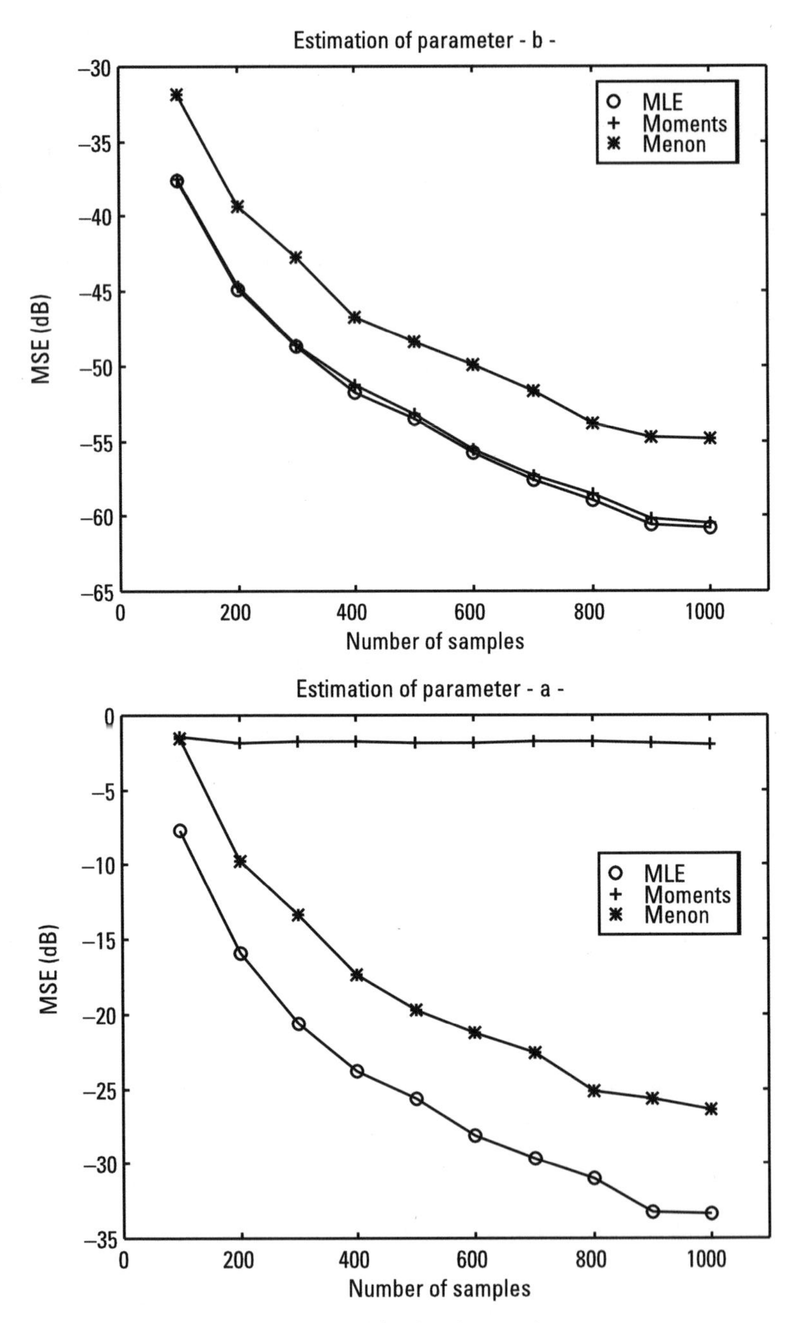


**Figure 6.7** MSE for the estimation of *a* and *b* using three estimators.

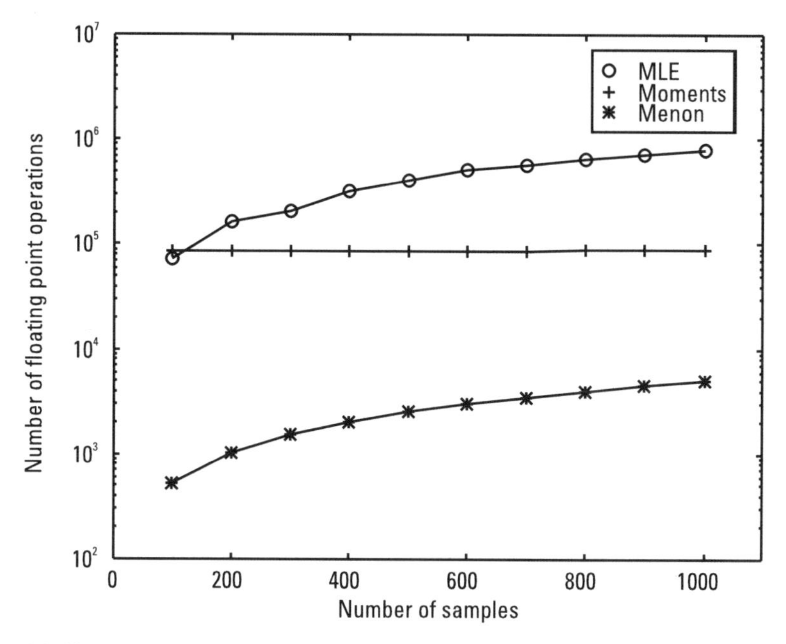


**Figure 6.8** Number of floating point operations used for each estimator.

$$\mu_{(j)} = m_{(j)} \quad \text{for} \quad j = 1, 2, \ldots, k \tag{6.9}$$

Because there are two unknown parameters, the mean and the second moment are used for the estimation so that two equations with two unknowns are formulated.

The theoretical firsts and second moments of the Weibull PDF are defined by

$$\mu_{(1)} = a^{-1/b}\Gamma\left(1 + \frac{1}{b}\right) \tag{6.10}$$

$$\mu_{(2)} = a^{-2/b}\Gamma\left(1 + \frac{2}{b}\right) \tag{6.11}$$

where $\Gamma$ denotes the gamma function defined by

$$\Gamma(r) = \int_0^\infty x^{r-1}e^{-x}dx \quad \text{for} \quad r > 0 \tag{6.12}$$

The sample first and second moments are defined by

$$m_{(1)} = \frac{1}{n}\sum_{i=1}^{n} x_i \tag{6.13}$$

and

$$m_{(2)} = \frac{1}{n}\sum_{i=1}^{n} x_i^2 \tag{6.14}$$

Solving for $b$ first:

$$\frac{\mu_{(1)}^2}{\mu_{(2)}} = \frac{m_{(1)}^2}{m_{(2)}} \tag{6.15}$$

or

$$\frac{\left[\Gamma\left(1 + \frac{1}{\hat{b}}\right)\right]^2}{\Gamma\left(1 + \frac{2}{\hat{b}}\right)} = \frac{\left[\frac{1}{n}\sum_{i=1}^{n} x_i\right]^2}{\frac{1}{n}\sum_{i=1}^{n} x_i^2} \tag{6.16}$$

To find $\hat{b}$, (6.16) has to be solved numerically. Subsequently, $\hat{a}$ can be found as

$$\hat{a} = \left[\frac{\Gamma\left(1 + \frac{1}{\hat{b}}\right)}{\frac{1}{n}\sum_{i=1}^{n} x_i}\right]^{\hat{b}} \tag{6.17}$$

It should be noted that no close expression exists for the estimation of $b$. Still, the solution of (6.16) and (6.17) requires fewer computations than would the MLE, as illustrated in Figure 6.8, at the expense of biasing $\hat{a}$, as shown in Figure 6.7.

### 6.2.4 Weibull Parameter Estimation Via Menon's Method

Menon [4] proposed two simple statistics for the estimation of the Weibull parameters. The two expressions are given by

$$\hat{b} = \left\{ \frac{6}{\pi^2} \frac{n}{n-1} \left[ \frac{1}{n} \sum_{i=1}^{n} (\ln x_i)^2 - \left( \frac{1}{n} \sum_{i=1}^{n} \ln x_i \right)^2 \right] \right\}^{-1/2} \tag{6.18}$$

and

$$\hat{a} = \left[ \exp\left( \frac{\hat{b}}{n} \sum_{i=1}^{n} \ln x_i + 0.577216 \right) \right]^{-1} \tag{6.19}$$

The solution of those expressions is based on the following transformation of the Weibull random variable $X$:

$$W = (X/a^{-1/b})^b \tag{6.20}$$

Finding the cumulative distribution function (CDF) of $W$, we have

$$F_W(w) = P(W < w) = P[(X/a^{-1/b})^b < w] = P(X < a^{-1/b} w^{1/b}) \tag{6.21}$$

The PDF is then found by taking the derivative of the CDF,

$$f_W(w) = \frac{d}{dw} F_X(a^{-1/b} w^{1/b}) = \exp(-w) I_{(0,\infty)}(w) \tag{6.22}$$

Note that the PDF of $W$ does not depend on $a$ or $b$. The variance of the logarithm of $W$ is given by

$$\mathrm{var}(\ln W) = \int_0^\infty (\ln w)^2 e^{-w} dw - \left[ \int_0^\infty (\ln w) e^{-e} dw \right]^2 = \frac{\pi^2}{6} \tag{6.23}$$

From the transformation in (6.20), the variance should also be equal to:

$$\mathrm{var}(\ln W) = \mathrm{var}\{[\ln(a^{1/b}) + b \ln X]\} = b^2\, \mathrm{var}(\ln X) \tag{6.24}$$

Equating those two results and using the sample moments for computing $\mathrm{var}(\ln X)$, the expression for $\hat{b}$ is obtained.

In order to estimate $\hat{a}$, the expected value of the logarithm of (6.20) is computed as

$$E(\ln W) = E[b \ln(a^{1/b} X)] = \ln a + bE[\ln X] = -0.577216 \tag{6.25}$$

Using the previous result for $\hat{b}$ as well as the sample moments, solving for $a$ yields the other parameter estimation statistic.

### 6.2.5 Example of Weibull Parameter Estimation

From the different methods presented so far, the method proposed by Menon seems to be the more practical in terms of computational time. A simulation is performed in order to choose one of the methods in which the computational number of operations and the error between the true and estimated parameters are considered.

To generate samples from a random variable distributed as Weibull, the theorem of stochastic simulation was used. Using the standard uniform distribution, samples from $X$ are obtained as follows [5].

Given the Weibull CDF of $X$

$$F_X(x) = 1 - \exp(-ax^b) \qquad \text{for } x \geq 0 \tag{6.26}$$

and having samples from the standard uniform distribution $U(0, 1)$ generated by a computer, the random variable

$$X = F_X^{-1}(u) \tag{6.27}$$

where

$$F_X^{-1}(u) = \inf[x : F_X(x) \geq u] \qquad \text{for } 0 \leq u \leq 1 \tag{6.28}$$

has a Weibull CDF.

Matlab® was used to obtain the results using the MLE approach. The method of moments was computed with the help of a table derived from the following expression [6]:

$$\Gamma(1 + z)\Gamma(1 + z)/\Gamma(1 + 2z) \qquad \text{for } 0 \leq z \leq 0.99 \tag{6.29}$$

Menon's method required just the implementation of the two statistics presented in the discussion of this method.

Next an array of samples generated by a Weibull distribution was generated. The length of this array varied from 100 up to 1,000 samples in steps of 100. For each array, 1,500 realizations were obtained, and the MSE for both parameter estimations were computed, as shown in Figure 6.7. In both graphs, the minimum error occurs using the MLE method. The $b$ parameter plot has very similar curves for the three estimators, which is not the case for $a$, where the method of moments shows that the estimation is biased and deviates considerably from the other two estimators. A second plot is shown in Figure 6.8, where the numbers of floating point operations for each estimator are shown. It can be seen that the optimum operator according to this variable is the one proposed by Menon. There is a considerable difference in terms of the number of computational operations needed between this approach and the other two. Thus, the method of choice in this work for estimating the Weibull parameters is Menon's method.

### 6.2.6 CFAR Filtering

The probability of false alarm (PFA) is defined as [7]

$$PFA = \int_{\eta}^{\infty} f_X(x)dx \tag{6.30}$$

so the decision threshold can be expressed as

$$\eta = \left(\frac{|\ln(PFA)|}{a}\right)^{1/b} \tag{6.31}$$

Simulations of CFAR filtering in the range domain were performed to verify the efficiency of the noise suppression scheme.

Gaussian distributed noise was generated and added to the signal of single scatterer. Range profiles were generated via the IDFT. The corresponding Weibull parameters and threshold were calculated by processing the response of 128 range-cells dominated by noise for each range profile. The parameters $a$ and $b$ were calculated to be 4,425 and 1.7251 respectively, and the threshold $\eta$ was set to 0.02361. In implementing CFAR filtering, any sample with a magnitude above the threshold level is, by hypothesis, the response of a target at range $r$. Thus, the width and position of bandpass filters are easily determined. Subsequently, an FIR filter can be implemented to filter the noise and smooth the target's frequency domain signature.

Figure 6.9 shows the range profile of a signal before and after CFAR filtering. Figure 6.9(a) shows the range profile of a target where the SNR is 0 dB. In Figure 6.9(b), noise is reduced by an FIR bandpass filter in the frequency domain, which is determined by setting the threshold to 0.02361 in the range domain. As shown, CFAR filtering effectively suppressed the noise.

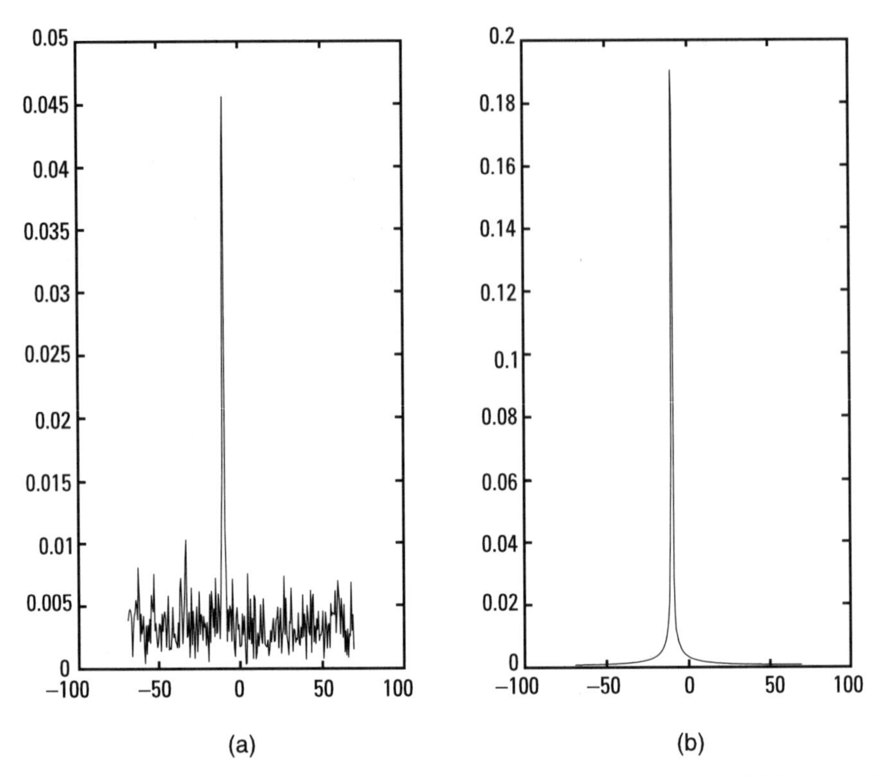


**Figure 6.9** (a) Range profile of a target showing the effects of additive noise; (b) FIR-filtered range profile for PFA = 0.001.

## 6.3 Complex Analysis of Point Scatterers

### 6.3.1 One Point Scatterer

Once a target signature is motion-compensated, the next objective is to determine if there is any residual range walk in the target's range profile history. With conventional range processing, one could use the peak intensity or strength of the range profile to track the target. For instance, consider a moving target initially located at $-10$m. Using conventional ISAR processing, the peak position is measured at $-9.86$m as shown in Figure 6.9(a).

Alternatively, complex analysis processing can be used to measure target position in the frequency domain. Complex analysis entails the use of the unwrapped phase slope of the target's echo transfer function to determine the target range. The unwrapping process corrects the phase angles in a vector by adding multiples of 2, when absolute jumps are greater than $\pi$.

Following that approach, the target's range is determined by multiplying the normalized phase slope by the actual range window size. For better accuracy, the phase slope is measured in regions where the noise interference is small. In Figure 6.10, the vertical lines indicate the limits of narrow intervals where the phase slope is to be measured. At an SNR level of 10 dB, the average target position measured through phase slope analysis is $-10.14$m. This suggests that phase slope analysis is comparable to conventional range domain processing for high SNR.

In an interval which noise is relatively weak, the amplitude of the signal will be high and the unwrapped phase will be nearly straight. In such an interval, an accurate target position measurement can be performed. Larger amplitude and phase variations worsen the precision in measurements. Increasing the SNR is imperative for precise phase slope measurements.

Figure 6.11 shows the amplitude and the phase of a frequency domain signature at a SNR of 0 dB. In this case, range estimation via complex analysis cannot be performed because the amplitude and the phase of the target's signature does not satisfy the minimum criteria for complex analysis. The condition must satisfy that the magnitude of the signal exceeds the threshold for certain

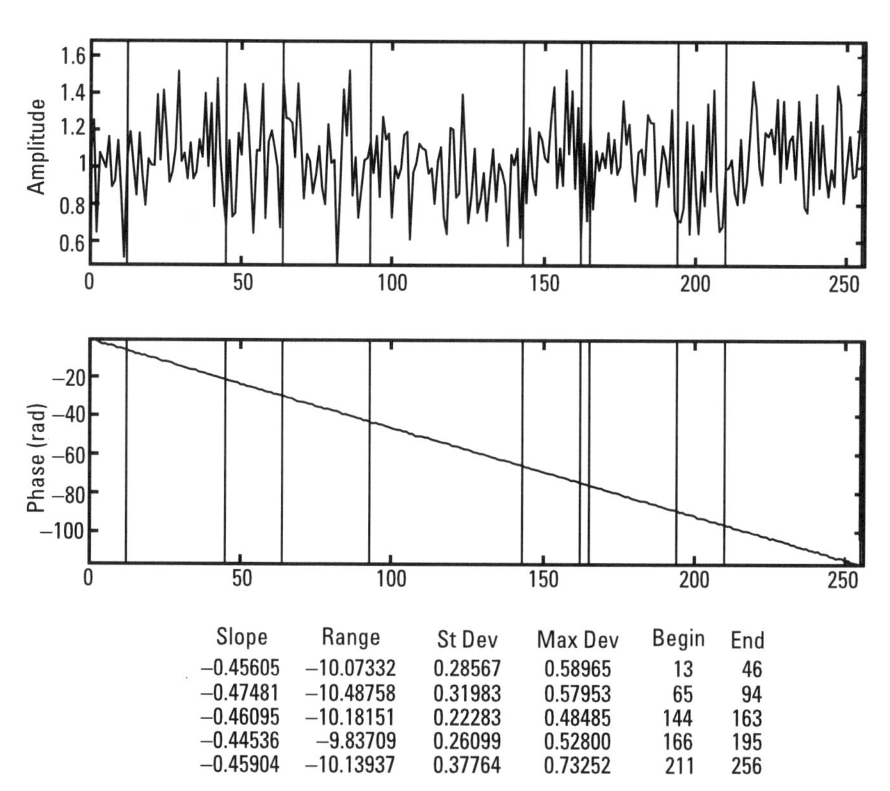


| Slope | Range | St Dev | Max Dev | Begin | End |
|---|---|---|---|---|---|
| −0.45605 | −10.07332 | 0.28567 | 0.58965 | 13 | 46 |
| −0.47481 | −10.48758 | 0.31983 | 0.57953 | 65 | 94 |
| −0.46095 | −10.18151 | 0.22283 | 0.48485 | 144 | 163 |
| −0.44536 | −9.83709 | 0.26099 | 0.52800 | 166 | 195 |
| −0.45904 | −10.13937 | 0.37764 | 0.73252 | 211 | 256 |

**Figure 6.10** Signal with SNR = 10 dB.

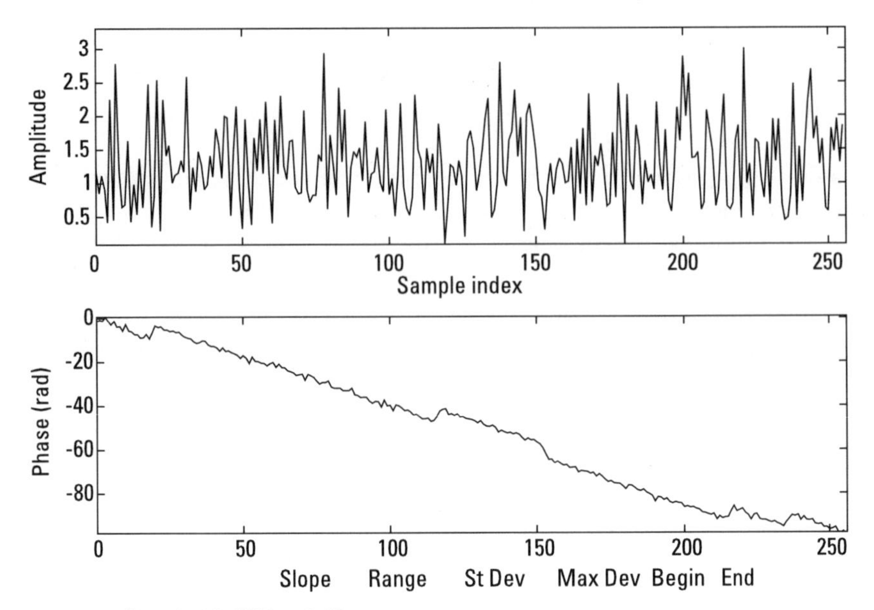


**Figure 6.11** Signal with SNR = 0 dB.

length and the phase does not deviate too much from the expected phase deviation. That condition can be met by improving the SNR using CFAR filtering and it ensures that the measured target position is accurate.

In this example, the SNR can be increased 3 dB by suppressing the noise in the range domain outside the region R(−45, 25) (m). The process yields the smoothed target signature amplitude and phase, as shown in Figure 6.12. Three segments are identified in which the amplitude is relatively high and the phase is nearly linear. Outside of the selected intervals, there is more than one occasional phase jump due to phase noise. Within the chosen intervals, the phase shows little deviation from a straight line. The result of additional filtering is illustrated in Figure 6.13. Here, an SNR of 10 dB is obtained by suppressing noise outside the region R(−17, −3) (m).

In the preceding example, the amplitude threshold was fixed to 70% of the mean amplitude of the burst. Likewise, the phase deviation threshold was fixed to 250% of the expected phase deviation, where the expected phase deviation is defined as [2, 8]

$$\sigma = \frac{1}{\sqrt{SNR}} \tag{6.32}$$

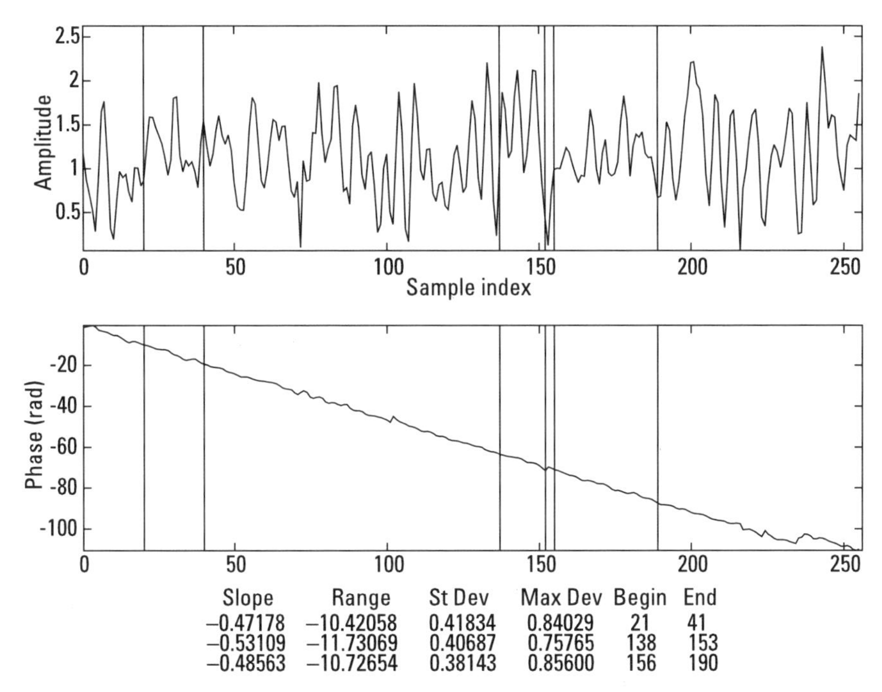


**Figure 6.12** Filtered signal with one-half the original noise power.

For comparison, Figure 6.14 illustrates the effect of CFAR filtering for a PFA of 0.001. When noise is suppressed using CFAR and FIR filtering, the amplitude fluctuation of the target's signature can become very small, and the unwrapped phase is quite straight. In addition, the interval in which the phase slope is measured spans the entire data set. The calculated target position is close to the actual position of the scatterer.

### 6.3.2 Two Point Scatterer Case

We have showed that complex analysis can be used successfully to measure the position of one point scatterer. In this section, we show that the same is true for two point scatterers. Figure 6.15 shows the amplitude and phase patterns for two fixed-point scatterers of equal strength separated by 1m. In this instance, the radar resolution is 0.54m, so that the two scatterers can be resolved. In the figure, one scatterer is located at −10.5m and the other scatterer is at −9.5m. The echo signal from the two point scatterers is

$$U = A \exp\left[i\left(\hat{\theta} - \frac{\phi}{2}\right)\right] + A \exp\left[i\left(\hat{\theta} + \frac{\phi}{2}\right)\right] \tag{6.33}$$

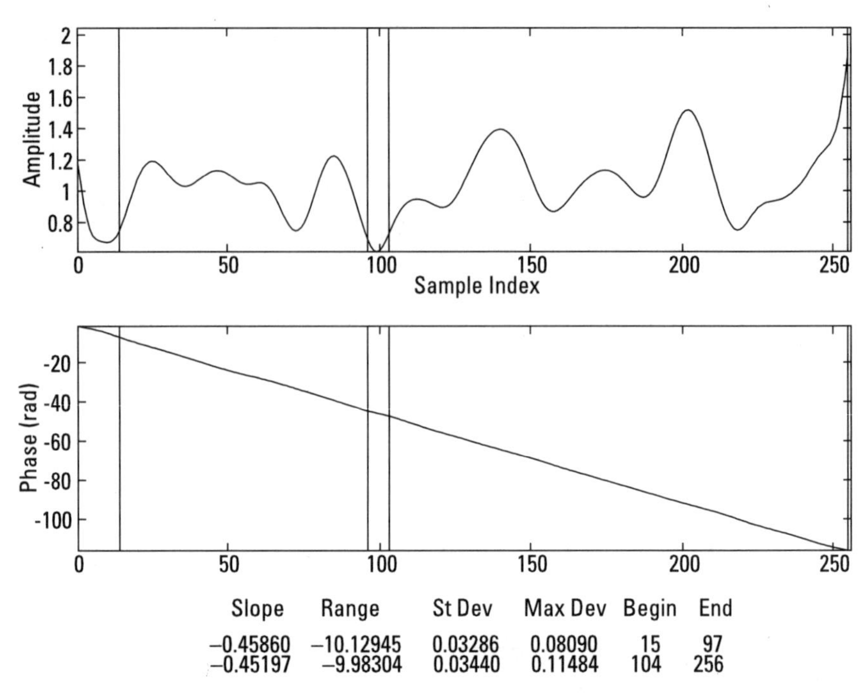


| Slope | Range | St Dev | Max Dev | Begin | End |
|---|---|---|---|---|---|
| −0.45860 | −10.12945 | 0.03286 | 0.08090 | 15 | 97 |
| −0.45197 | −9.98304 | 0.03440 | 0.11484 | 104 | 256 |

**Figure 6.13** Filtered signal with one tenth the original noise power.

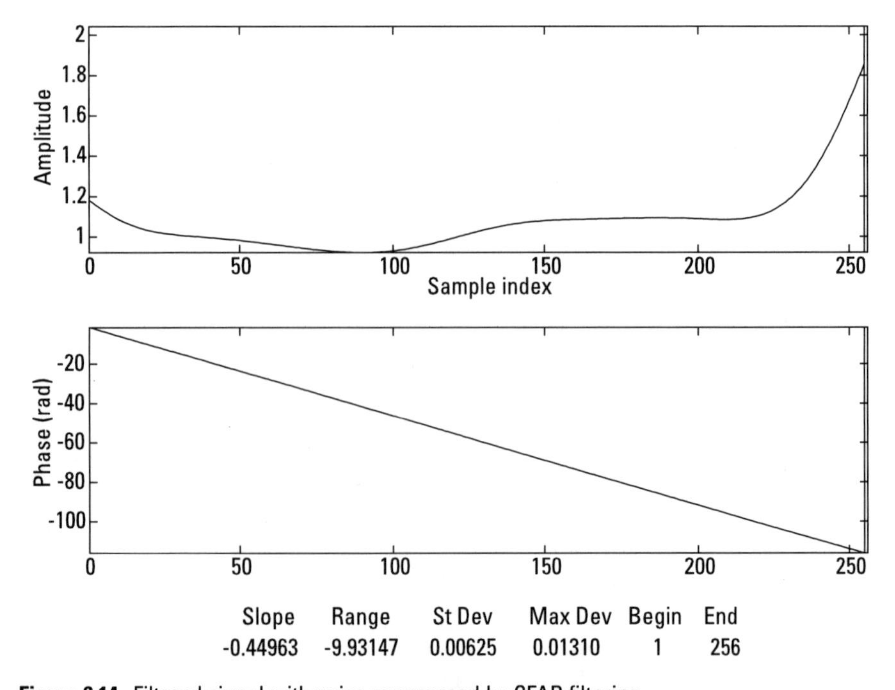


| Slope | Range | St Dev | Max Dev | Begin | End |
|---|---|---|---|---|---|
| -0.44963 | -9.93147 | 0.00625 | 0.01310 | 1 | 256 |

**Figure 6.14** Filtered signal with noise suppressed by CFAR filtering.

where $\theta$ represents the range to the center of the target and $\phi$ represents the range difference between the two scatterers. Notice that (6.33) can be written as

$$U = 2A\cos\left(\frac{\phi}{2}\right)\exp(i\hat{\theta}) \tag{6.34}$$

It follows that, unlike the case of one point scatterer case, the amplitude fluctuates, as illustrated in Figure 6.15.

However, the same procedure of complex analysis for the case of one point scatterer can be applied to the two point scatterer return signal. Figure 6.16 shows the amplitude and phase of the frequency domain signature for an SNR of 0 dB. Using only one segment, it yields an inaccurate range estimate. When the noise is suppressed using CFAR and FIR filtering, as shown in Figure 6.17, the amplitude and the phase of the target's signature are comparable to the infinite SNR case. The range estimate calculated via phase slope analysis is close to the actual centroid value of the two point scatterers.

As discussed in [8, 9], the advantage of using the intensity and the phase information of a signal is evident when the received signal components are the reflection of a complex man-made target.

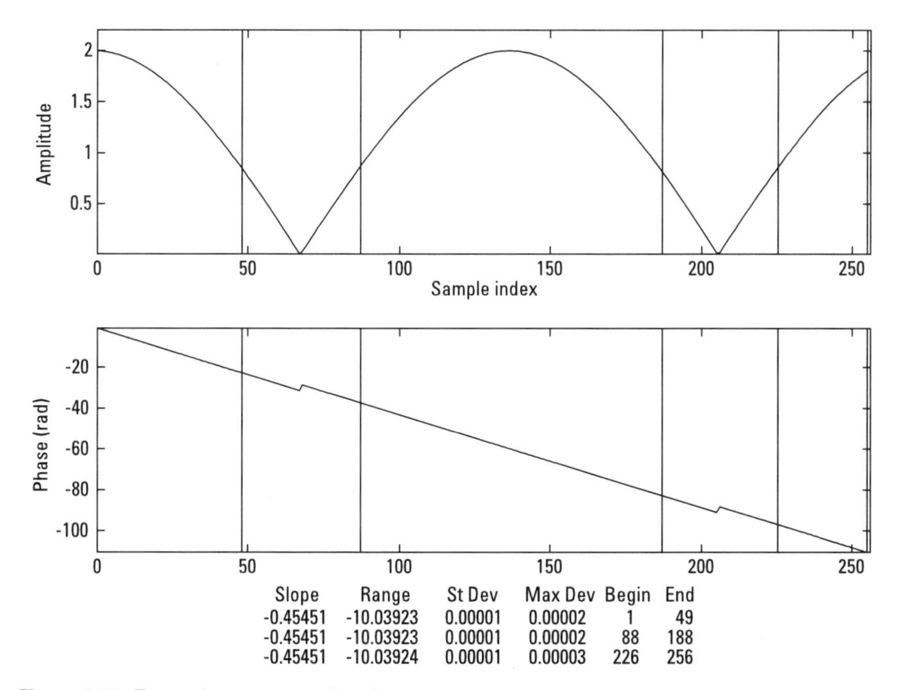


**Figure 6.15** Two point scatterer signal.

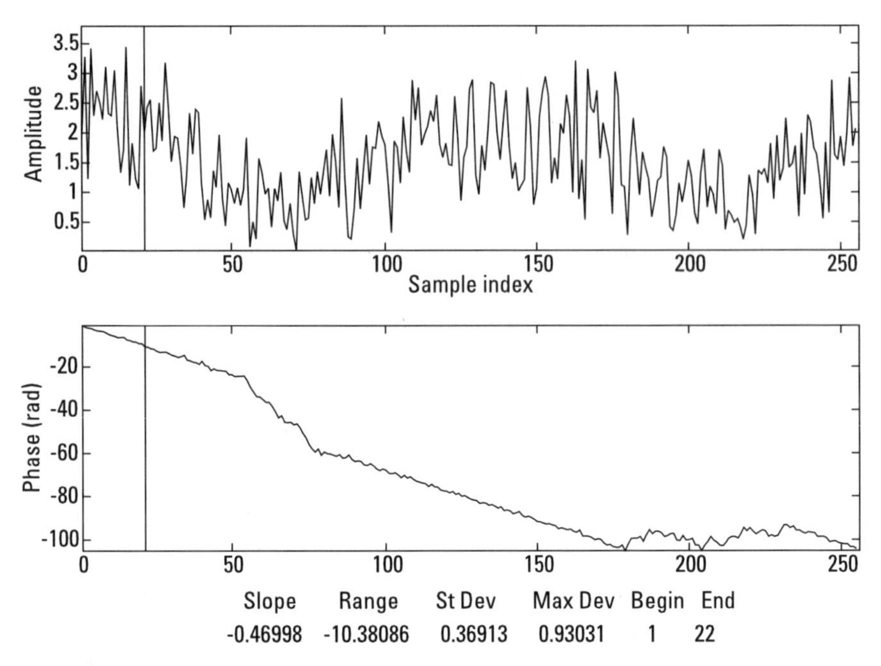


| Slope | Range | St Dev | Max Dev | Begin | End |
|---|---|---|---|---|---|
| -0.46998 | -10.38086 | 0.36913 | 0.93031 | 1 | 22 |

**Figure 6.16** Two point scatterer signal with SNR = 0 dB.

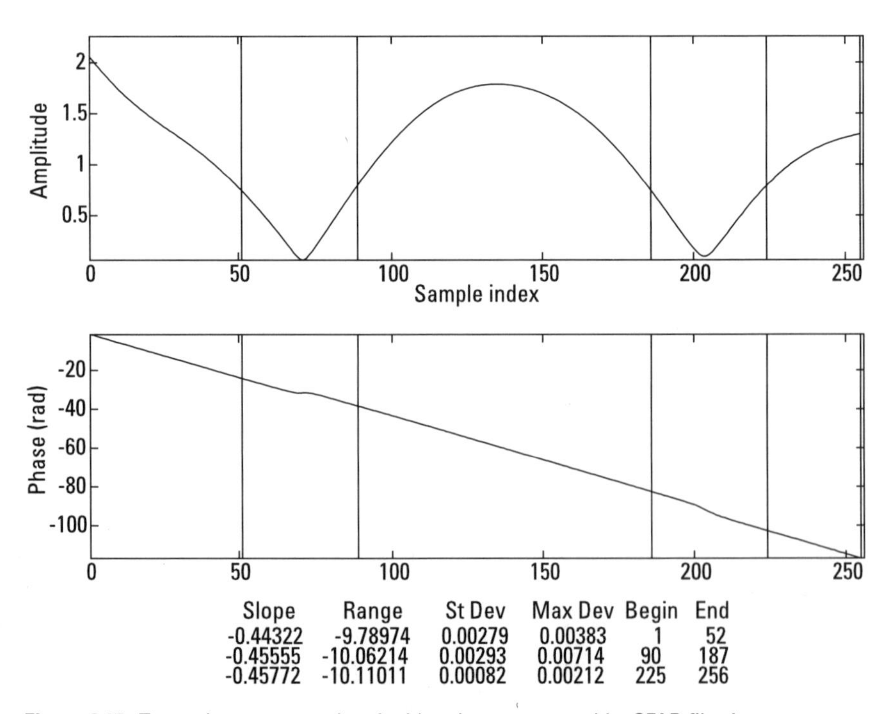


| Slope | Range | St Dev | Max Dev | Begin | End |
|---|---|---|---|---|---|
| -0.44322 | -9.78974 | 0.00279 | 0.00383 | 1 | 52 |
| -0.45555 | -10.06214 | 0.00293 | 0.00714 | 90 | 187 |
| -0.45772 | -10.11011 | 0.00082 | 0.00212 | 225 | 256 |

**Figure 6.17** Two point scatterer signal with noise suppressed by CFAR filtering.

## 6.4 Results for a Man-Made Target

The proposed complex analysis approach was applied to a Boeing 727 signature burst. Figure 6.18 shows the amplitude and the phase of that signature. Amplitude fluctuations and phase discontinuities are obvious. When the complex analysis is performed on the unprocessed signature, there is no good section of data that can be processed to determine the target range.

Figure 6.19 shows the effect of CFAR and FIR filtering. Here, the amplitude fluctuations are significantly reduced and the phase is well behaved. In this case, four segments of the signature are chosen to determine the instantaneous target range. The calculated target range spans from 4.4345m to 3.6851m, decreasing consistently from segment to segment, which shows that the target's centroid is moving toward the radar.

## 6.5 Summary

This chapter demonstrated that complex analysis can enhance target range estimation via phase slope measurements. It showed that the high SNR needed

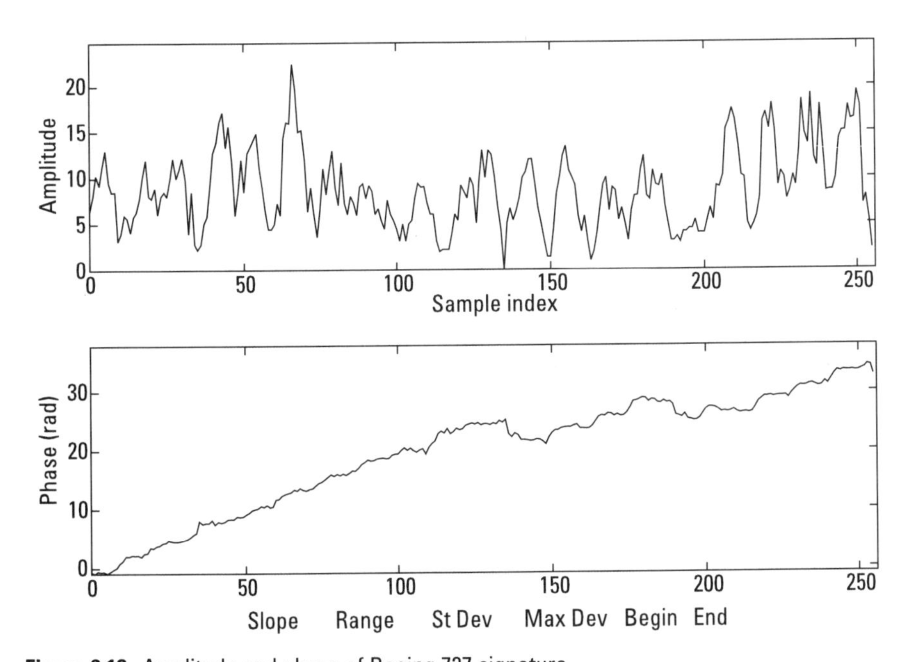


**Figure 6.18** Amplitude and phase of Boeing 727 signature.

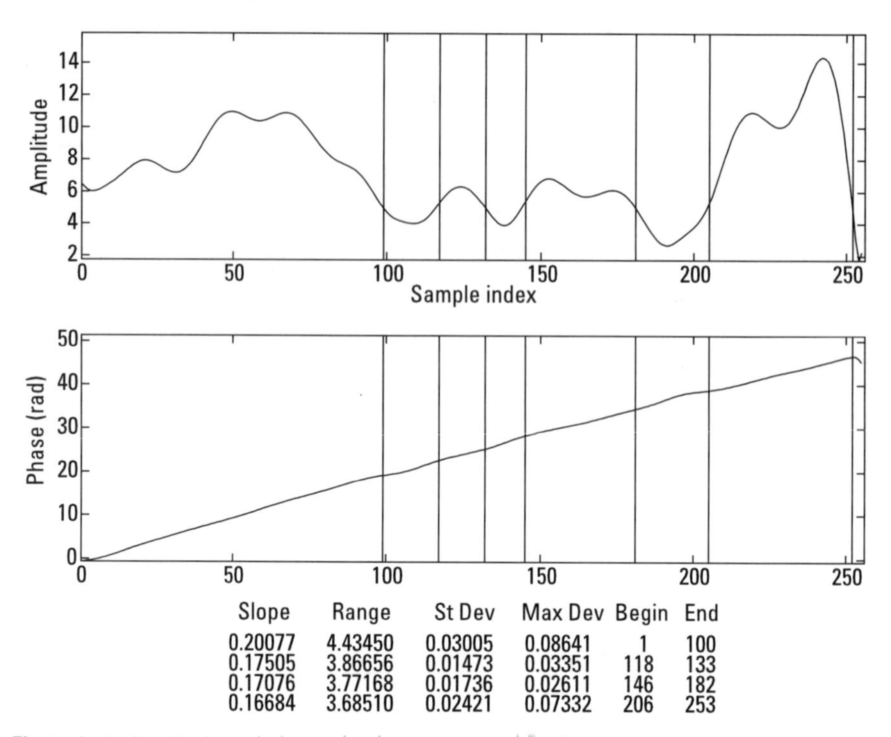


| Slope | Range | St Dev | Max Dev | Begin | End |
|---|---|---|---|---|---|
| 0.20077 | 4.43450 | 0.03005 | 0.08641 | 1 | 100 |
| 0.17505 | 3.86656 | 0.01473 | 0.03351 | 118 | 133 |
| 0.17076 | 3.77168 | 0.01736 | 0.02611 | 146 | 182 |
| 0.16684 | 3.68510 | 0.02421 | 0.07332 | 206 | 253 |

**Figure 6.19** Amplitude and phase of noise suppressed Boeing 727 signature.

for precise phase slope measurements could be achieved through CFAR filtering by modeling noise according to the Weibull distribution and subsequent FIR filtering. CFAR and subsequent FIR filtering on the support regions have the following effects:

- Amplitude fluctuation reduction;
- Phase linearity;
- Segment length increase.

Range measurement simulations were performed using point scatterers. Results demonstrated that complex analysis can be used to determine the range accurately so that the target estimation based on these range measurements will be accurate. In the case of an actual target signature, the use of complex analysis yielded consistent range measurements. The only outstanding issue in using the complex analysis is that the range estimation performance is difficult to predict because the number of segments and the number of samples in a segment may vary from burst to burst.

# References

[1] Sekine, M., and Y. Mao, *Weibull Radar Clutter*, London: Peter Peregrinus, 1990.

[2] Levanon, N., *Radar Principles*, New York: Wiley.

[3] Larsen, R. J., and M. L. Marx, *An Introduction to Mathematical Statistics and Its Applications*, 2nd ed., Englewood Cliffs, NJ: Prentice Hall, 1986.

[4] Menon, M. V., "Estimation of the Shape and Scale Parameters of the Weibull Distribution," *Technometrics*, Vol. 5, No. 2, May 1963, pp. 175–182.

[5] Maisel, H., and G. Gnugnoli, *Simulation of Discrete Stochastic Systems*, Science Research Associates, 1972.

[6] Olkin, I., L. J. Gleser, and C. Derman, *Probability Models and Applications*, Macmillan, 1980.

[7] Whalen, A. D., *Detection of Signals in Noise*, San Diego: Academic Press, 1971.

[8] Rihaczek A. W., and S. J. Hershkowitz, *Radar Resolution and Complex-Image Analysis*, Norwood, MA: Artech House, 1996.

[9] Rihaczek, A. W., and S. J. Hershkowitz, "Man-Made Target Backscattering Behavior: Applicability of Conventional Radar Resolution Theory," *IEEE Trans. on Aerospace and Electronic Systems*, Vol. 32, April 1996.

# 7

# Cramer-Rao Bounds for Stepped-Frequency Waveform

Chapter 6 discussed the complex analysis of ISAR signatures. Although the proposed technique is accurate, an exact target range cannot be provided because noise cannot be completely eliminated from the signature. This chapter examines a method to assess the accuracy of motion parameter estimation using phase slope measurements. The method entails the calculation of Cramer-Rao bounds for corresponding motion estimates.

The Cramer-Rao bound is the lowest theoretical MSE achievable in parameter estimation. It holds regardless of the method used to obtain the parameter estimation, provided the estimator is unbiased. In other words, the Cramer-Rao lower bound dictates that the MSE corresponding to the estimator of a parameter cannot be smaller than a certain quantity related to the likelihood function. If an estimator's variance is equal to the Cramer-Rao lower bound, then such an estimator is called efficient. The Cramer-Rao lower bounds for motion estimation using a stepped-frequency waveform are derived in this chapter. Here, Monte Carlo simulations are performed to test the validity of those Cramer-Rao bounds.

## 7.1 Cramer-Rao Bound for Range Estimation

Using the duality theorem of Fourier integrals, the following relationships between time domain and frequency domain signals have been established [1]:

- The range delay $\tau$ is the dual of the frequency $f$.
- The effective time duration $t_d$ is the dual of the effective bandwidth $\beta_e$.

Recall that the frequency Cramer-Rao bound for an unbiased estimator [2] is

$$\sigma_f^2 \geq \frac{1}{4\pi^2 \frac{2E}{N_0} t_d^2} \tag{7.1}$$

where $t_d^2$ is the second moment about the signal's mean time and $2E/N_0$ is the output peak SNR.

Consequently, the Cramer-Rao variance bound for the range delay estimation is written as

$$\sigma_\tau^2 \geq \frac{1}{4\pi^2 \frac{2E}{N_0} \beta_e^2} \tag{7.2}$$

where $\beta_e$ is the effective bandwidth of the signal. The relation between the target range $\tau$ and its range delay are proportional to each other as

$$r = \frac{c}{2}\tau \tag{7.3}$$

Thus, the range standard deviation is given by

$$\sigma_r = \frac{c}{2}\sigma_r \tag{7.4}$$

From (7.2) and (7.4), it follows that the Cramer-Rao variance bound for range estimation is

$$\sigma_R^2 \geq \left(\frac{c}{2}\right)^2 \frac{1}{4\pi^2 \dfrac{2E}{N_0} \beta_e^2} \tag{7.5}$$

Because the slant-range resolution $\Delta r$ is inversely proportional to the synthetic bandwidth $2\beta_e$ for the stepped-frequency waveform radar, and $2E/N_0$ is the SNR, then (7.5) can be written in terms of slant range and SNR as [2]

$$\sigma_R^2 \geq \frac{3\Delta r^2}{\pi^2 SNR} \tag{7.6}$$

The Cramer-Rao bound is derived next for range estimation via complex analysis. Assume that the $k$th segment of the echo transfer function has $\tilde{N}_k$ samples. Then the slant range resolution changes to $\Delta \tilde{r}_k = (N/\tilde{N}_k)\Delta r$. From (7.6), it follows that

$$\sigma_R^2 \geq \frac{3\Delta \tilde{r}_k^2}{\pi^2 SNR_k} = \frac{3N^2 \Delta r^2}{\tilde{N}_k^2 SNR_k} \tag{7.7}$$

where $\text{SNR}_k$ denotes the SNR associated with the $k$th segment of the target's signature.

Note that the Cramer-Rao bound is inversely related with $SNR$, which indicates that the range measurement accuracy can be improved with higher $SNR$. Also note that the Cramer-Rao bound is minimum when $N = \tilde{N}_k$, which is all the data in a burst used for the range estimation.

## 7.2 Cramer-Rao Bound for Velocity Estimation

According to [2], the variance of the velocity is given by

$$\sigma_v^2 \geq \frac{12\sigma_r^2}{LT^2} \tag{7.8}$$

where $L$ is the total number of segments to be used and $T$ is integration time. Using (7.7) in (7.8) yields

$$\sigma_v^2 \geq \frac{36 N^2 \Delta r^2}{L \tilde{N}_k^2 \pi^2 T^2 SNR_k} \tag{7.9}$$

Equation (7.9) is the Cramer-Rao bound for velocity estimation when the number of samples and amplitudes in a selected segment are the same for all the selected segments. See Appendix C for a detailed derivation of (7.9). The special case of an equal number of samples in segments is $\tilde{N}_k = N$ for all $k$ and $L = M$. In other words, the number of bursts equals the number of signature segments, and the number of pulses per segment equals the number of samples in a burst. In this case, the Cramer-Rao velocity estimation bound reduces to:

$$\sigma_v^2 \geq \frac{36\Delta r^2}{M\pi^2 T^2 SNR} \tag{7.10}$$

Next, we determine how the velocity Cramer-Rao bound is affected when a weighted least squares approach is used to estimate range.

Whereas the least squares method minimizes the sum of the squares of the error $\mathbf{e}^t\mathbf{e}$, the weighted least squares method minimizes the sum of the squares of the error weighted by $\mathbf{W}$, that is, $\mathbf{e}^t\mathbf{W}\mathbf{e}$. In the case of the weighted least squares method being used for the motion estimation, the measurement error $\mathbf{e}$ to be equivalent to the velocity standard deviation $\sigma_v$, or acceleration standard deviation $\sigma_a$, so that the minimization of $\mathbf{e}^t\mathbf{W}\mathbf{e}$ is equivalent to minimizing $\sigma_\mathbf{v}^t\mathbf{W}\sigma_\mathbf{v}$ or $\sigma_\mathbf{a}^t\mathbf{W}\sigma_\mathbf{a}$.

From the orthonormal relation of the velocity variance $\sigma_\mathbf{v}^2$ with respect to the weighting function $\mathbf{W}_\mathbf{v}$, we have

$$\sigma_{vk}^2 w_{vk}^2 = 1 \tag{7.11}$$

The weighting matrix for velocity motion coefficients $\mathbf{W}_v$ is derived by multiplying the weighting matrix $\mathbf{W}$ by a constant $1/K_v$. Thus, that relation can be written as

$$\mathbf{W}_\mathbf{v} = \frac{\mathbf{W}}{K_v} \tag{7.12}$$

From (7.11) and (7.12), the nonzero element $w_{vk}^2$ can be written as

$$w_{vk}^2 = \frac{w_k}{K_v} \tag{7.13}$$

Then $\sigma_{vk}^2 w_k^2 = \mathrm{K_v}$, which is a constant for every $k$. See Chapter 8 for more details.

Define the velocity variance vector $\sigma_v^2$ as

$$\sigma_v^2 = [\sigma_{v1}^2, \sigma_{v2}^2, \ldots, \sigma_{vL}^2 \tag{7.14}$$

and the weighting matrix **W** as

$$\mathbf{W} = \begin{bmatrix} w_1^2 & 0 & \cdots & 0 \\ 0 & w_2^2 & \cdots & 0 \\ \vdots & \vdots & \ddots & \vdots \\ 0 & 0 & \cdots & w_L^2 \end{bmatrix} \tag{7.15}$$

From (7.8), we have

$$\sigma_{vk}^2 \geq \frac{36N^2 \Delta r^2}{\tilde{N}_k^2 \pi^2 T^2 SNR_k} \tag{7.16}$$

Because $SNR_k$ is proportional to the average amplitude in the $k$th segment, $\sigma_{vk}^2$ can be written as

$$\sigma_{vk}^2 \geq \frac{36N^2 \Delta r^2}{\tilde{N}_k^2 \pi^2 T^2 \left(\frac{\mu_k}{\mu}\right)^2 SNR} \tag{7.17}$$

where

$$\mu_k = \sum_{j=1}^{\tilde{N}_k} \frac{|A_{kj}|}{N_k} \tag{7.18}$$

$$\mu = \frac{\left(\sum_{j=1}^{MN} |A_{kj}|\right)}{MN} \tag{7.19}$$

$$SNR_k = \left(\frac{\mu_k}{\mu}\right)^2 SNR \tag{7.20}$$

Now, define $w_k$ as

$$w_k = \tilde{N}_k \mu_k = \sum_{j=1}^{\tilde{N}_k} |A_{kj}| \tag{7.21}$$

which is a sum of sample magnitudes in the $k$th segment. Thus, $\sigma_{vk}^2$ in (7.17) can be expressed in terms of $w_k$ as

$$\sigma_{vk}^2 \geq \frac{K_v}{w_k^2} \tag{7.22}$$

where

$$K_v = \frac{36 N^2 \Delta r^2 \mu^2}{\pi^2 T^2 SNR} \tag{7.23}$$

From (7.21), it is obvious that the total weight can be written as

$$w_T^2 = w_1^2 + w_2^2 + \cdots + w_L^2 = \sum_{k=1}^{L} w_k^2 \tag{7.24}$$

From (7.22), it follows that $\sigma_{vk}^2$ and $w_k^2$ are the reciprocal of each other. Thus, the total velocity variance is written as

$$\frac{1}{\sigma_v^2} = \frac{1}{\sigma_{v1}^2} + \frac{1}{\sigma_{v2}^2} + \cdots + \frac{1}{\sigma_{vL}^2} = \sum_{k=1}^{L} \frac{1}{\sigma_{vk}^2} \tag{7.25}$$

or

$$\sigma_v^2 = \left( \sum_{k=1}^{L} \frac{1}{\sigma_{vk}^2} \right)^{-1} \tag{7.26}$$

where

$$\sigma_{v1}^2 w_1^2 = \sigma_{v2}^2 w_2^2 = \cdots = \sigma_{vk}^2 w_k^2 = \text{constant} \tag{7.27}$$

By substituting (7.22) into (7.26), the total velocity variance can be expressed as

$$\sigma_v^2 \geq \left[ \sum_{k=1}^{L} \frac{SNR\pi^2 T^2 \left( \sum_{j=1}^{\tilde{N}_k} |A_{kj}| \right)^2}{36N^2 \Delta r^2 \left( \sum_{j=1}^{MN} |A_j| \right)^2 / (MN)^2} \right]^{-1} \tag{7.28}$$

Thus, the total velocity standard deviation is

$$\sigma_v \geq \left[ \sum_{k=1}^{L} \frac{SNR\pi^2 T^2 \left( \sum_{j=1}^{\tilde{N}_k} |A_{kj}| \right)^2}{36N^2 \Delta r^2 \left( \sum_{j=1}^{MN} |A_j| \right)^2 / (MN)^2} \right]^{-1/2} \tag{7.29}$$

Note that the Cramer-Rao bound depends not only on the number of bursts, number of pulses in a burst, SNR, or integration time but also on the magnitude of the signal when the weighted least squares method is used for the motion estimation.

Figure 7.1 shows the behavior of the Cramer-Rao bound for velocity estimation given in (7.29). Here, the Cramer-Rao bound is a function of the *SNR* and the number of bursts. The number of samples in a burst is set to 256 throughout the processing. The integration time $T$ is 0.7585, 1.5170, 3.0341, and 6.0681 sec for 16, 32, 64, and 128 bursts, respectively. All other radar system parameters are fixed to the values given in Appendix A. As the number of bursts processed increases, the Cramer-Rao velocity estimation bound decreases.

Figure 7.2 displays the Cramer-Rao bound for velocity estimation as a function of the number of samples and the SNR using (7.29). The number of bursts is constant at 64. $T$ is 0.3793, 0.7585, 1.5170, and 3.0341 sec for 32, 64, 128, and 256 pulses in a burst. Notice that the velocity standard deviation decreases as the number of frequency samples increases.

Figure 7.3 displays the Cramer-Rao velocity estimation bound as a function of the number of pulses, number of intervals, and the SNR. In this case, the integration time $T$ is fixed to 3.0341 sec. The number of segments is 8, 4, 2, and 1 per burst for 32, 64, 128, and 256 pulses per segment. Here, the Cramer-Rao bound for velocity estimation is lower when there are more pulses per segment.

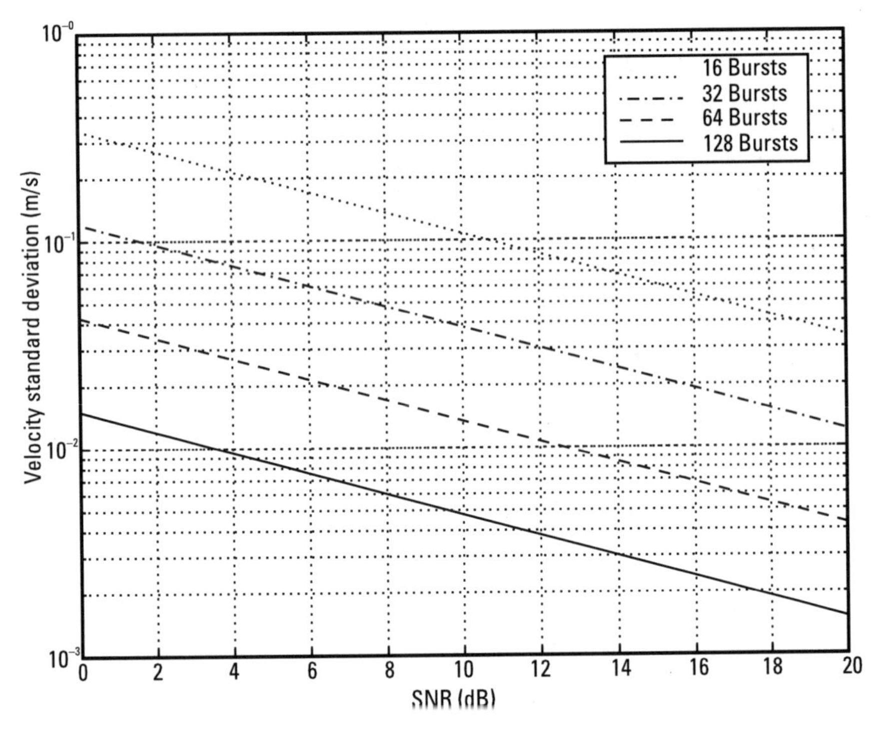


**Figure 7.1** Cramer-Rao bound for velocity estimation as a function of the number of bursts.

## 7.3 Cramer-Rao Bound for Acceleration Estimation

The same procedure used to find the Cramer-Rao bound for velocity estimation can be applied to the Cramer-Rao bound for acceleration estimation. According to Levanon [2], the variance of an unbiased acceleration estimator is

$$\sigma_a^2 \geq \frac{720\sigma_r^2}{LT^4} \tag{7.30}$$

Using (7.7) in (7.30), we obtain

$$\sigma_a^2 \geq \frac{2160N^2\Delta r^2}{L\tilde{N}_k^2\pi^2T^4 SNR} \tag{7.31}$$

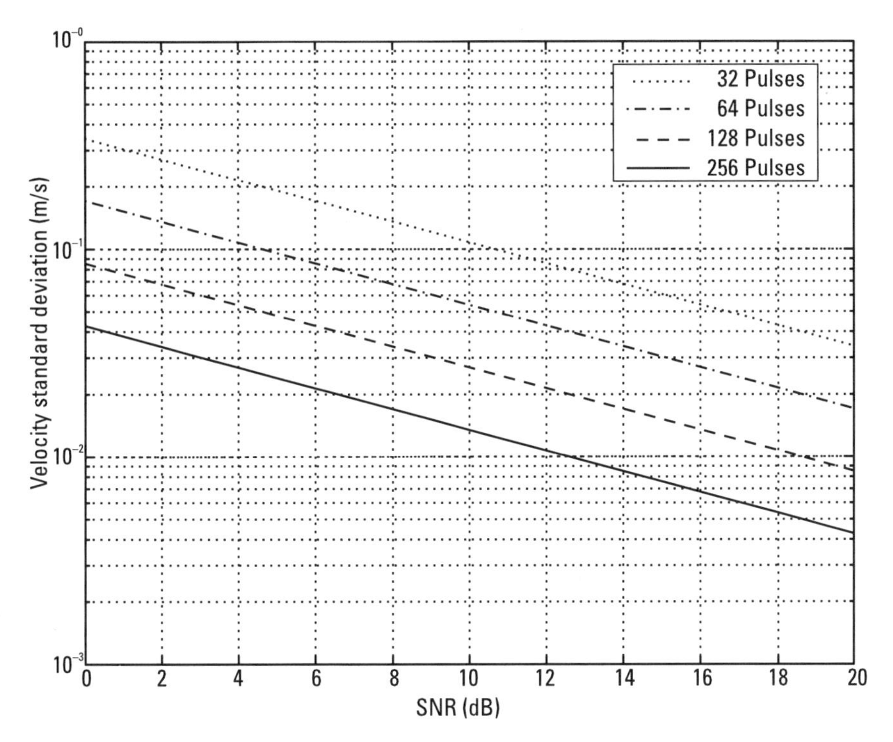


**Figure 7.2** Cramer-Rao bound for velocity estimation as a function of the number of frequency samples.

When a complete data set is used for motion estimation, the Cramer-Rao bound for acceleration estimation reduces to

$$\sigma_a^2 \geq \frac{2160\Delta r^2}{M\pi^2 T^4 SNR} \tag{7.32}$$

Define the acceleration variance vector $\sigma_{\mathbf{a}}^2$ as

$$\sigma_{\mathbf{a}}^2 = [\sigma_{a1}^2, \sigma_{a2}^2, \ldots, \sigma_{aL}^2] \tag{7.33}$$

From (7.30), we have

$$\sigma_{ak}^2 \geq \frac{2160 N^2 \Delta r^2}{\tilde{N}_k^2 \pi^2 T^4 SNR} \tag{7.34}$$

or

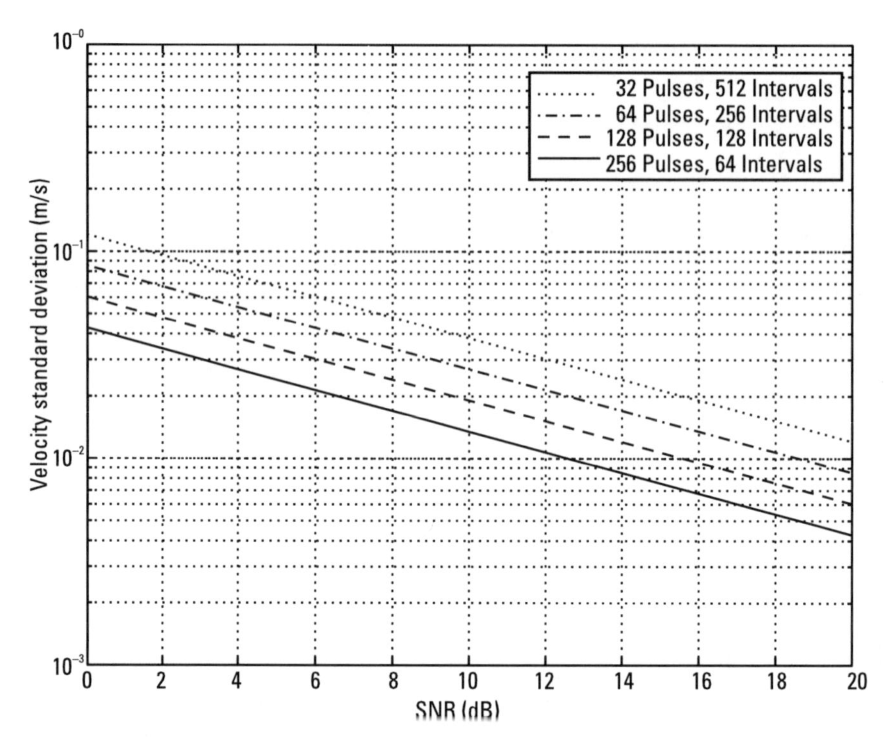


**Figure 7.3** Cramer-Rao bound for velocity estimation as a function of the numbers of frequency samples and intervals.

$$\sigma_{ak}^2 \geq \frac{2160 N^2 \Delta r^2}{\tilde{N}_k^2 \pi^2 T^4 \left(\frac{\mu_k}{\mu}\right)^2 SNR} \tag{7.35}$$

Further simplification of (7.35) leads to

$$\sigma_{ak}^2 = \frac{K_a}{w_k^2} \tag{7.36}$$

where

$$K_a = \frac{2160 N^2 \Delta r^2 \mu^2}{\pi^2 T^4 SNR} \tag{7.37}$$

Now, recall that

$$w_T^2 = w_1^2 + w_2^2 + \cdots + w_L^2 = \sum_{k=1}^{L} w_k^2 \tag{7.38}$$

Because $\sigma_{ak}^2$ and $w_k^2$ are the reciprocal of each other, the total acceleration variance can be written as

$$\frac{1}{\sigma_a^2} = \frac{1}{\sigma_{a1}^2} + \frac{1}{\sigma_{a2}^2} + \cdots + \frac{1}{\sigma_{aL}^2} = \sum_{k=1}^{L} \frac{1}{\sigma_{ak}^2} \tag{7.39}$$

or

$$\sigma_a^2 = \left( \sum_{k=1}^{L} \frac{1}{\sigma_{ak}^2} \right)^{-1} \tag{7.40}$$

Substituting (7.36) into (7.37) yields

$$\sigma_a^2 = \left[ \sum_{k=1}^{L} \frac{SNR\pi^2 T^4 \left( \sum_{j=1}^{\tilde{N}_k} |A_{kj}| \right)^2}{2160 N^2 \Delta r^2 \left( \sum_{j=1}^{MN} |A_j| \right)^2 /(MN)^2} \right]^{-1} \tag{7.41}$$

Thus, the acceleration standard deviation is

$$\sigma_a = \left[ \sum_{k=1}^{L} \frac{SNR\pi^2 T^4 \left( \sum_{j=1}^{\tilde{N}_k} |A_{kj}| \right)^2}{2160 N^2 \Delta r^2 \left( \sum_{j=1}^{MN} |A_j| \right)^2 /(MN)^2} \right]^{-1/2} \tag{7.42}$$

Like the Cramer-Rao bound for velocity, the bound for acceleration depends on the magnitude of the signal when the weighted least squares method is used for the motion estimation.

Figure 7.4 illustrates the behavior of the Cramer-Rao bound for acceleration estimation as given by (7.42). In the figure, the Cramer-Rao bound is a function of the SNR and the number of bursts. The number of samples in a burst is kept constant at 256 throughout the processing. The integration time $T$ is 0.7585, 1.5170, 3.0341, and 6.0681 sec for 16, 32, 64, and 128 bursts, respectively. All other radar system parameters are fixed to the values given in Appendix A. Observe that the Cramer-Rao bound for acceleration estimation decreases as the number of bursts processed increases.

Figure 7.5 displays the Cramer-Rao bound for acceleration estimation as a function of the number of samples and the SNR. The number of bursts is set constant to 64. $T$ is 0.3793, 0.7585, 1.5170, and 3.0341 sec for 32, 64, 128, and 256 pulses in a burst. As the number of frequency samples increases, the acceleration standard deviation decreases.

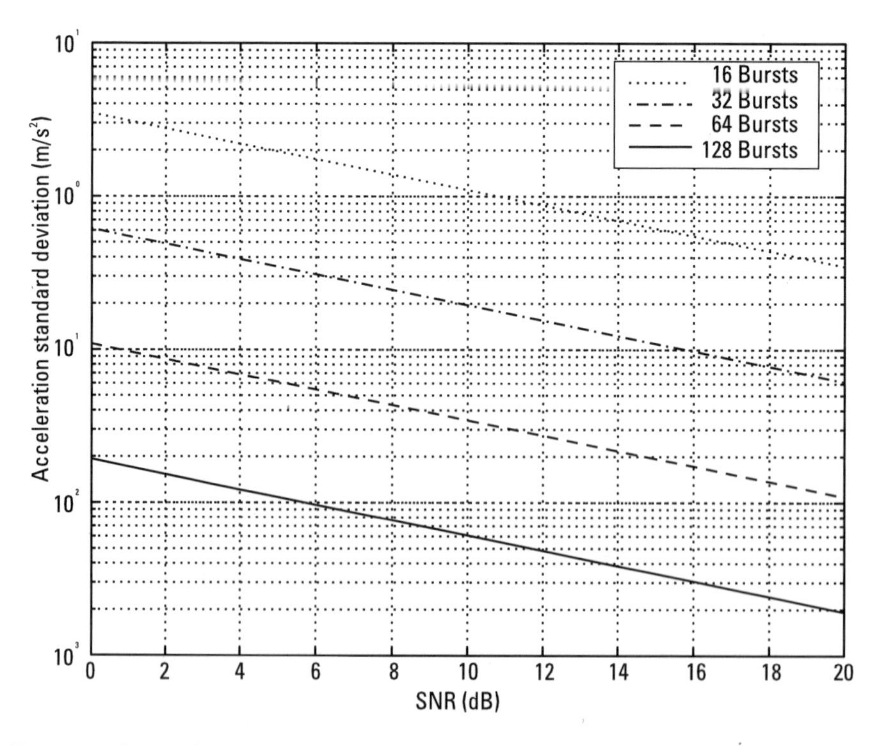


**Figure 7.4** Cramer-Rao bound for acceleration estimation as a function of the number of bursts.

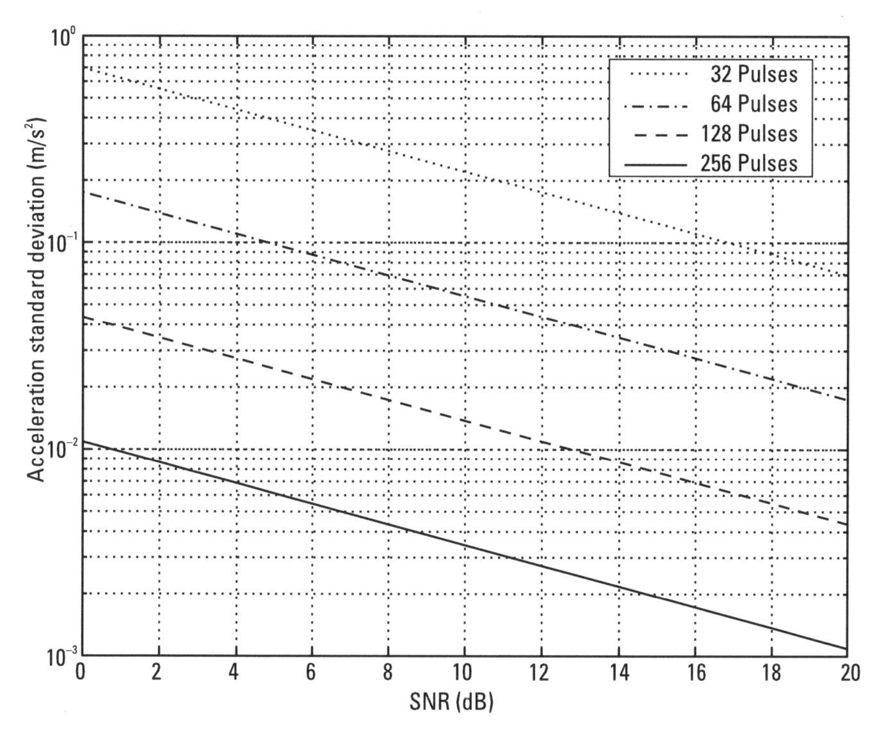


**Figure 7.5** Cramer-Rao bound for acceleration estimation as a function of the number of frequency samples.

Figure 7.6 displays the Cramer-Rao acceleration estimation bound as a function of the number of pulses, the number of intervals, and the SNR. In this case, the integration time $T$ is fixed to 3.0341 sec. The number of intervals is 8, 4, 2, and 1 per burst for 32, 64, 128, and 256 pulses per segment. It is clear that the Cramer-Rao acceleration estimation bound is lower when there are more pulses in a segment.

## 7.4 Monte Carlo Simulations

This section describes the experimental results obtained via Monte Carlo simulations. Figures 7.7 and 7.8 display the velocity and acceleration Cramer-Rao bounds as a function of the SNR. The solid lines correspond to the velocity and

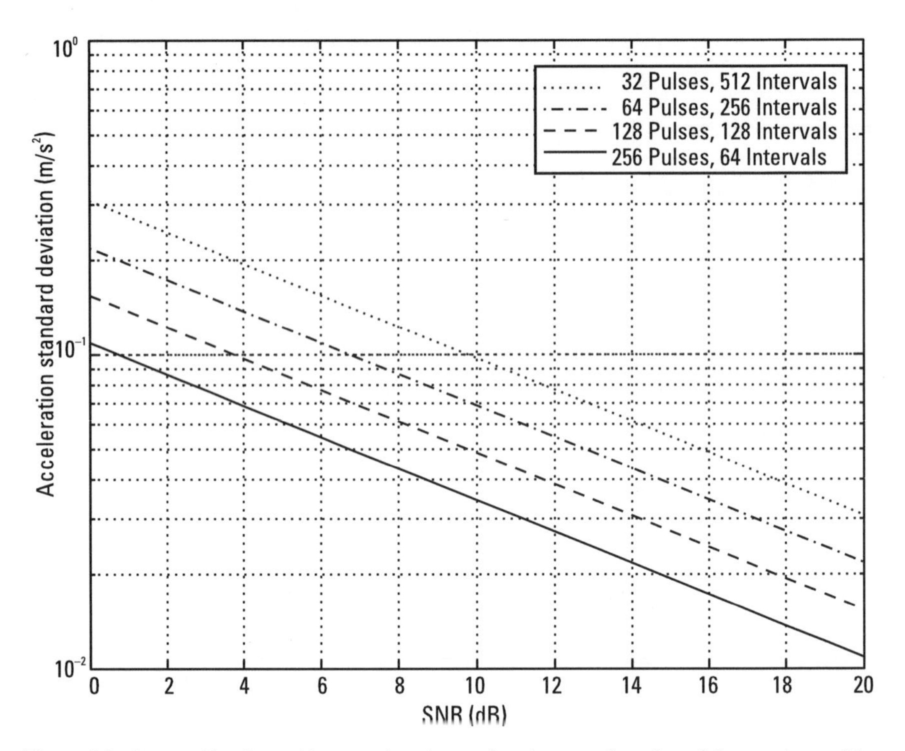


**Figure 7.6** Cramer-Rao bound for acceleration estimation as a function of the numbers of frequency samples and intervals.

acceleration standard deviation obtained via simulations. The dashed lines represent the Cramer-Rao bound using the entire data set. Equation (7.10) and (7.32) are represented by the dashed lines in Figures 7.7 and 7.8, respectively. The number of Monte Carlo simulations was 100. The velocity and acceleration granularity were set to 0.1 $\sigma_v$ and 0.1 $\sigma_a$. The SNR was increased from 0 dB to 20 dB in 5-dB increments. Table 7.1 shows the details of the simulation parameters.

The velocity and acceleration Cramer-Rao bounds that account for a weighted least squares–based complex analysis closely match experimental results. That is particularly true for high SNR and velocity and acceleration variances equal to or greater than the Cramer-Rao bound.

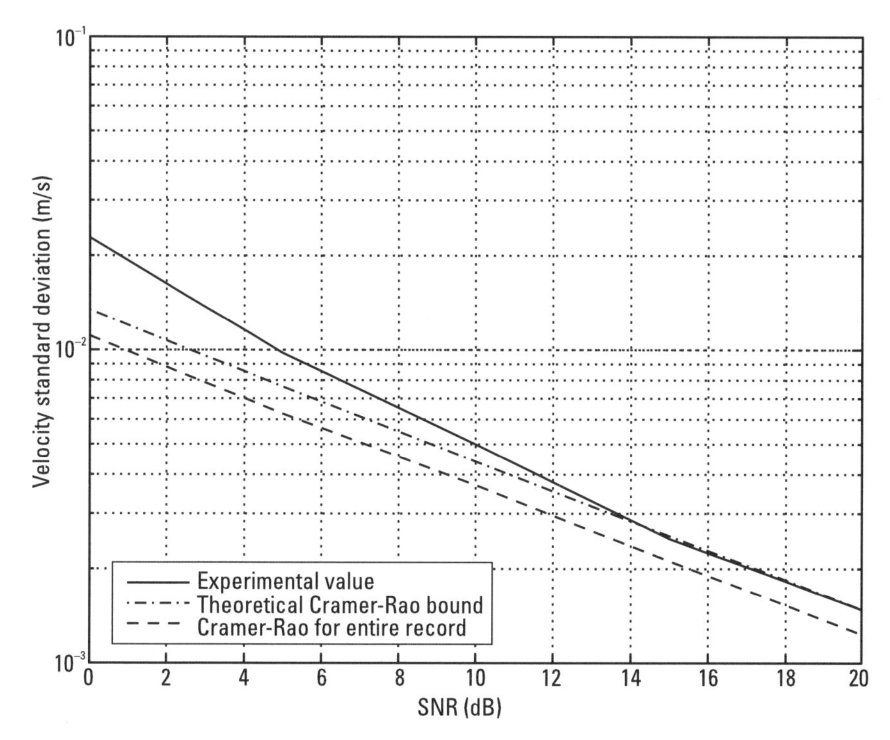


**Figure 7.7** Monte Carlo simulations for velocity standard deviation.

**Table 7.1**
Simulation Parameters Used in Determining the Standard Deviation of Target Motion Estimates

| Parameter | Symbol | Value |
|---|---|---|
| Target position | $r_0$ | −10.5m, −9.5m |
| Target velocity | $v_0$ | −10 m/s |
| Target acceleration | $a_0$ | 5 m/s$^2$ |
| Number of Monte Carlo simulation | $K_{mc}$ | 100 |
| Velocity granularity | $\delta_v$ | 0.1 m/s |
| Acceleration granularity | $\delta_a$ | 0.1 m/s$^2$ |
| Signal to noise ratio | SNR | 0, 5, 10, 15, 20 dB |

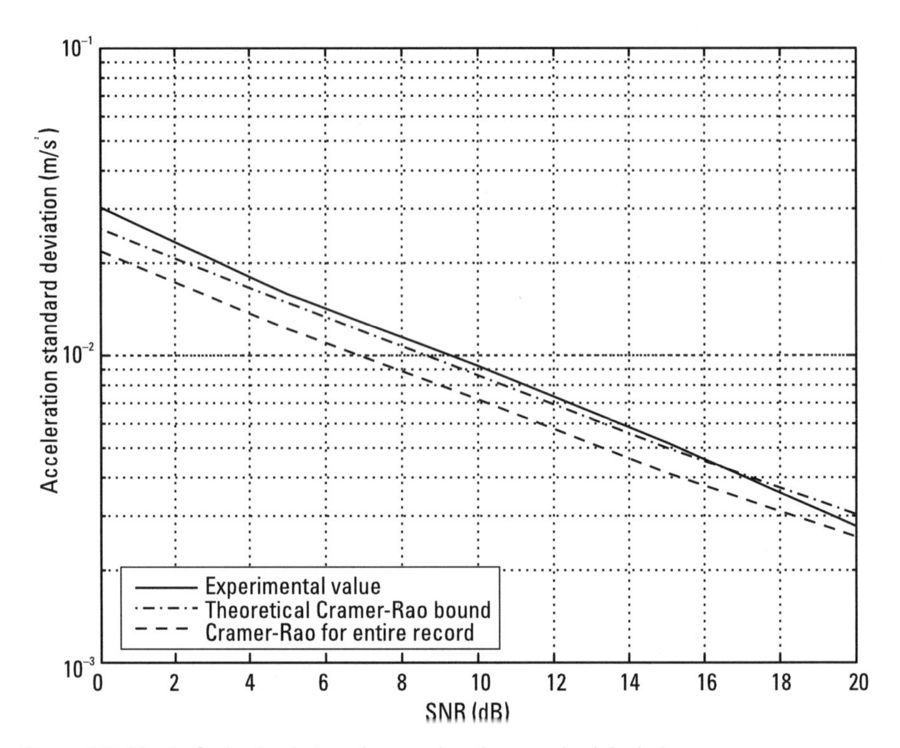


**Figure 7.8** Monte Carlo simulations for acceleration standard deviation.

## 7.5 Summary

Cramer-Rao bounds for the stepped-frequency waveform were derived to assess the accuracy achievable in ISAR motion estimation using phase slope measurements. Specifically, the Cramer-Rao bounds for range, velocity, and acceleration were determined when the target motion is estimated using the weighted least squares method. Monte Carlo simulations were performed to verify the validity of the velocity and acceleration Cramer-Rao bounds.

## References


[1] Martinez, A., *Parameter Estimation Theory for Synthetic High Resolution Radar*, Master's thesis, Univ. of Texas at El Paso, July 1998.

[2] Levanon, N., *Radar Principles*, New York: Wiley, 1988.

# 8

# Weighted Least Squares Motion Parameter Estimation

This chapter describes a new method for improved motion estimation that assigns weights to the segments of the echo transfer function used for instantaneous range estimation. The method is suitable for motion estimation from signatures with varying amplitude and number of samples per segment. For instance, a typical target at microwave frequencies can be thought of as consisting of multiple scatterers. The magnitude of the phasor sum of the echoes from individual backscattering centers will fluctuate rapidly in time, even for small aspect changes of the target. Slight changes in the radar frequency also changes the relative phase between backscattering centers and thereby produce changes in the narrowband echo magnitude. Therefore, the magnitude of echo signal from the target in high-resolution radar always fluctuates. The magnitude fluctuation affects SNR, and it is well known that high SNR is desired for better motion estimation. Usually SNR is high when the magnitude is large. Using weighted least squares method for motion parameter estimation, greater weights are assigned to segments of the target's signature with large amplitude where SNR is high so that the segment can provide more accurate phase information.

## 8.1 Least Squares for Unevenly Spaced Data

Upon determining the instantaneous position of the target, we proceed to estimate the motion parameters from a set of overdetermined equations. The change of the instantaneous position provides the information about target kinematics. The minimum number of data needed to estimate the target velocity and acceleration is three, if the data are noise free. However, the radar signal is usually corrupted by many factors and cannot be considered in practice as a noise-free signal. In a noise-corrupted signal, more data than are minimally required provide better parameter estimation. If more data than required are used for the estimation, it becomes an over-determined equation and is written as

$$\mathbf{A}\mathbf{x} = \mathbf{b} \tag{8.1}$$

where $\mathbf{A}$ is an $m \times n$ matrix with $m > n$ and elements are instantaneous time index when the target range is measured. In (8.1), the estimates of the motion parameters determined in a least square sense are given by the elements of $\hat{\mathbf{x}}$ while the range information is assigned to $\mathbf{b}$. For consistency, Eq. (8.1) is reformulated as

$$\mathbf{A}\hat{\mathbf{x}} + \mathbf{e} = \mathbf{b} \tag{8.2}$$

where $\hat{\mathbf{x}}$ is an estimate of $\mathbf{x}$ and $\mathbf{e} = \mathbf{A}(\mathbf{x} - \hat{\mathbf{x}})$ is the difference between the measured value and the estimated value. The best approximation to $\hat{\mathbf{x}}$ is obtained when the error vector $\mathbf{e}$ is minimal. That is commonly done by solving the fundamental least squares problem given as

$$\mathbf{J}(\hat{\mathbf{x}}) = (\mathbf{b} - \mathbf{A}\hat{\mathbf{x}})^{\mathbf{b}}(\mathbf{b} - \mathbf{A}\hat{\mathbf{x}}) = \mathbf{e}^{\mathbf{t}}\mathbf{e} \tag{8.3}$$

where the superscript $\mathbf{t}$ denotes the transpose of the matrix. The vector that minimizes (8.3) is

$$\hat{\mathbf{x}} = (\mathbf{A}^{\mathbf{t}}\mathbf{A})^{-1}\mathbf{A}^{\mathbf{t}}\mathbf{b} \tag{8.4}$$

Consider fitting a set of instantaneous range measurements to a $p$th-order polynomial. That is, let

$$r = \alpha_0 + \alpha_1 t + \alpha_2 t^2 + \cdots + \alpha_p t^p \tag{8.5}$$

Then

$$\hat{\mathbf{x}} = \begin{bmatrix} \alpha_0 \\ \alpha_1 \\ \vdots \\ \alpha_p \end{bmatrix} \tag{8.6}$$

and

$$\mathbf{A} = \begin{bmatrix} 1 & A_1 & A_1^2 & \cdots & A_1^p \\ 1 & A_2 & A_2^2 & \cdots & A_2^p \\ \vdots & \vdots & \vdots & \vdots & \vdots \\ 1 & A_L & A_L^2 & \cdots & A_L^p \end{bmatrix} \tag{8.7}$$

Since time elapses when sampling the echo signal within a segment, the elements in **A** are time coordinates for the middle of a segment. In (8.7), $A_k$ is the center coordinate of the $k$th segment of the signature, and subscript $L$ denotes the total number of chosen segments. The vector **b** contains the target position determined via phase slope measurements within the same intervals.

Because the width of the segments and the average amplitude within each segment vary from segment to segment, a weighted least squares method that minimizes

$$\mathbf{J}(\hat{\mathbf{x}}) = (\mathbf{b} - \mathbf{A}\hat{\mathbf{x}})^t\mathbf{W}(\mathbf{b} - \mathbf{A}\hat{\mathbf{x}}) = \mathbf{e}^t\mathbf{W}\mathbf{e} \tag{8.8}$$

becomes more appealing. In (8.8), **W** is a symmetric, positive, definite weighting matrix. The next section describes an approach for determining the weighting matrix.

## 8.2 Weighting Matrix

Consider a $1 \times L$ row vector

$$\mathbf{X} = (X_1, X_2, \ldots, X_L) \tag{8.9}$$

where $X_1, X_2, \ldots, X_L$ are random variables. The covariance matrix of **X** is an $L \times L$ matrix defined as

$$\Sigma_{\mathbf{x}} = \begin{bmatrix} \sigma_{x1x1} & \sigma_{x1x2} & \cdots & \sigma_{x1xL} \\ \sigma_{x2x1} & \sigma_{x2x2} & \cdots & \sigma_{x2xL} \\ \vdots & \vdots & \vdots & \vdots \\ \sigma_{xLx1} & \sigma_{xLx2} & \cdots & \sigma_{xLxL} \end{bmatrix} \tag{8.10}$$

Note that $\sigma_{xixj}$ is the covariance of $X_iX_j$. Assuming that the components of vector **X** are independent,

$$\sigma_{xixj} = \sigma_{ij} = 0, \quad i \neq j \tag{8.11}$$

so that **x** is a diagonal matrix given by

$$\Sigma_{\mathbf{x}} = \begin{bmatrix} \sigma_1^2 & 0 & \cdots & 0 \\ 0 & \sigma_2^2 & \cdots & 0 \\ \vdots & \vdots & \vdots & \vdots \\ 0 & 0 & \cdots & \sigma_L^2 \end{bmatrix} \tag{8.12}$$

If the random signals are zero-mean with a known positive covariance matrix, then the weighting matrix $\mathbf{W} = \Sigma_{\mathbf{x}}^{-1}$. Thus, the weighting matrix is expressed as

$$\mathbf{W} = \begin{bmatrix} \frac{1}{\sigma_1^2} & 0 & \cdots & 0 \\ 0 & \frac{1}{\sigma_2^2} & \cdots & 0 \\ \vdots & \vdots & \vdots & \vdots \\ 0 & 0 & \cdots & \frac{1}{\sigma_L^2} \end{bmatrix} \tag{8.13}$$

In ISAR signal processing, there is an interest in obtaining velocity and acceleration estimates. Thus, the corresponding weighting matrices are written in terms of velocity variance and acceleration variance as

$$\mathbf{W_v} = \begin{bmatrix} \frac{1}{\sigma_{v1}^2} & 0 & 0 \\ 0 & \frac{1}{\sigma_{v2}^2} & 0 \\ \vdots & \vdots & \vdots \\ 0 & 0 & \frac{1}{\sigma_{vL}^2} \end{bmatrix} \tag{8.14}$$

and

$$\mathbf{W_a} = \begin{bmatrix} \frac{1}{\sigma_{a1}^2} & 0 & \cdots & 0 \\ 0 & \frac{1}{\sigma_{a2}^2} & \cdots & 0 \\ \vdots & \vdots & \vdots & \vdots \\ 0 & 0 & \cdots & \frac{1}{\sigma_{aL}^2} \end{bmatrix} \tag{8.15}$$

The elements of matrices in (8.14) and (8.15) are given by standard deviations of the velocity and acceleration estimates associated with each signature segment. From Chapter 7, the variances of the velocity and acceleration motion coefficients for the $k$th segment are given as

$$\sigma_{vk}^2 = \frac{36N^2\Delta r^2\mu^2}{\tilde{N}_k^2\pi^2T^2\mu_k^2 SNR} \tag{8.16}$$

and

$$\sigma_{ak}^2 = \frac{2160N^2\Delta r^2\mu^2}{\tilde{N}_k^2\pi^2T^4\mu_k^2 SNR_k} \tag{8.17}$$

where subscript $k$ denotes the $k$th segment. Equations (8.16) and (8.17) can be simplified as

$$\sigma_{vk}^2 = \frac{K_v}{w_k^2} \tag{8.18}$$

and

$$\sigma_{ak}^2 = \frac{K_a}{w_k^2} \tag{8.19}$$

where $w_k = \tilde{N}_k\mu_k$ as defined in (7.21). $K_v$ and $K_a$ are defined as

$$K_v = \frac{36N^2\Delta r^2\mu^2}{\pi^2T^2 SNR} \tag{8.20}$$

and

$$K_a = \frac{2160N^2\Delta r^2\mu^2}{\pi^2T^4 SNR_k} \tag{8.21}$$

Substituting (8.14) and (8.15) into (8.18) and (8.19), respectively, the weighting matrices become

$$\mathbf{W_v} = \frac{1}{K_v}\begin{bmatrix} w_1^2 & 0 & \cdots & 0 \\ 0 & w_2^2 & \cdots & 0 \\ \vdots & \vdots & \vdots & \vdots \\ 0 & 0 & \cdots & w_L^2 \end{bmatrix} \tag{8.22}$$

and

$$\mathbf{W_a} = \frac{1}{K_a}\begin{bmatrix} w_1^2 & 0 & \cdots & 0 \\ 0 & w_2^2 & \cdots & 0 \\ \vdots & \vdots & \vdots & \vdots \\ 0 & 0 & \cdots & w_L^2 \end{bmatrix} \tag{8.23}$$

Because $K_v$ and $K_a$ are constants, they have no effect on the estimation problem and can be omitted. Therefore, the weighting matrix for the weighted least squares is simply

$$\mathbf{W} = \begin{bmatrix} w_1^2 & 0 & \cdots & 0 \\ 0 & w_2^2 & \cdots & 0 \\ \vdots & \vdots & \vdots & \vdots \\ 0 & 0 & \cdots & w_L^2 \end{bmatrix} \tag{8.24}$$

Elements of the weighting matrix are found by summing the magnitude in each segment, assuming the noise variance and magnitude in a signal are inversely related.

## 8.3 Weighted Least Squares Fitting

When we have more confidence in the accuracy of some of the phase slope measurements, we can choose the elements of $\mathbf{W}$ to weigh these measurements more heavily [1]. Expressing the positive definite weighting matrix $\mathbf{W}$ in terms of its square root as

$$\mathbf{W} = \Psi^t\Psi \tag{8.25}$$

(8.8) becomes

$$\mathbf{J}(\hat{\mathbf{x}}) = \Psi\mathbf{b} - (\Psi\mathbf{A})\hat{\mathbf{x}}]^t[\Psi\mathbf{b} - (\Psi\mathbf{A})\hat{\mathbf{x}}] \tag{8.26}$$

Equation (8.26) is similar to (8.3). Thus, the vector $\hat{\mathbf{x}}$ that minimizes (8.26) can be found by replacing $\mathbf{b}$ with $\Psi\mathbf{b}$ and $\mathbf{A}$ with $\Psi\mathbf{A}$ in (8.4). That yields

$$\hat{\mathbf{x}} = [(\mathbf{A}^t\Psi^t)(\Psi\mathbf{A})]^{-1}(\mathbf{A}^t\Psi^t)(\Psi\mathbf{b}) \tag{8.27}$$

or

$$\hat{\mathbf{x}} = (\mathbf{A}^t\mathbf{W}\mathbf{A})^{-1}\mathbf{A}^t\mathbf{W}\mathbf{b} \tag{8.28}$$

If the target motion is modeled as a second-order polynomial, the components of $\hat{\mathbf{x}}$ in (8.28) contain information on the initial range, velocity, and acceleration of the target.

As an example, consider the data set illustrated in Figure 6.19. In that example, let us estimate the initial velocity of the target. Each element in $\mathbf{A}$ is given by the average time index of the segment divided by the PRF. The velocity can be estimated using a first-order polynomial. Hence, $\mathbf{A}$ is a matrix which has two columns:

$$\mathbf{A} = \begin{bmatrix} 1 & 0.094 \\ 1 & 0.0232 \\ 1 & 0.0304 \\ 1 & 0.0425 \end{bmatrix} \tag{8.29}$$

From the instantaneous range measurements, we obtain

$$\mathbf{b} = \begin{bmatrix} 4.43450 \\ 3.86656 \\ 3.77168 \\ 3.68510 \end{bmatrix} \tag{8.30}$$

The weighting matrix is

$$\mathbf{W} = \begin{bmatrix} 790^2 & 0 & 0 & 0 \\ 0 & 88^2 & 0 & 0 \\ 0 & 0 & 219^2 & 0 \\ 0 & 0 & 0 & 480^2 \end{bmatrix} \tag{8.31}$$

Substituting (8.29), (8.30) and (8.31) into (8.4) yields

$$\hat{\mathbf{x}} = \begin{bmatrix} 4.6463 \\ -23.2198 \end{bmatrix} \tag{8.32}$$

Thus, the velocity estimate for this case is −23.2198 m/s. The first solution, 4.6463, is an estimated initial range.

## 8.4 Weighted Least Squares Focal Quality Indicator

In ISAR, target motion is usually modeled as a second order polynomial. Consequently, the measured instantaneous target position for the $k$th segment is

$$b_k = \frac{1}{2}\Delta a t_k^2 + \Delta v t_k + C \tag{8.33}$$

where $C$ is a constant. Our objective is to estimate the target motion parameters. Hence we adopt a second order polynomial model:

$$r = \alpha_0 + \alpha_1 t + \alpha_2 t^2 \tag{8.34}$$

We estimate $\alpha_1$ and $\alpha_2$ via the weighted least squares method so that $\alpha_2 = \Delta a/2$ and $\alpha_1 = \Delta v$. At this point, we define a new focal quality indicator as

$$\gamma = [\alpha_1^2 + (2\alpha_2)^2]^{1/2} = [(\Delta v)^2 + (\Delta a)^2]^{1/2} \tag{8.35}$$

It is easy to show that the focal quality indicator given in (8.35) will approach its minimum value when the actual values of velocity and acceleration are used to compensate the target's echo transfer function. As shown in (8.35), the focal quality is a two-variable function:

$$\gamma = f(\Delta v, \Delta a) \tag{8.36}$$

Let B = $\Delta v$ and D = $2\Delta a$ in (8.35). Then,

$$\gamma = (\mathrm{B}^2 + \mathrm{D}^2)^{1/2} \tag{8.37}$$

Let us say that the conditions for a global minimum given by

$$\frac{\partial \gamma}{\partial \mathrm{B}} = 0, \quad \frac{\partial \gamma}{\partial \mathrm{D}} = 0 \tag{8.38}$$

are satisfied by the coordinates ($\mathrm{B}_0$, $\mathrm{D}_0$). Performing the partial differentiation of (8.37) with respect to B and D yields

$$\frac{\partial \gamma}{\partial B} = \frac{\mathrm{B}}{(\mathrm{B}^2 + \mathrm{D}^2)^{1/2}}, \quad \frac{\partial \gamma}{\partial \mathrm{D}} = \frac{\mathrm{D}}{(\mathrm{B}^2 + \mathrm{D}^2)^{1/2}} \tag{8.39}$$

Clearly, the values that satisfy (8.38) are B = 0 and D = 0. In other words, a minimum is reached only when $\Delta v = \Delta a = 0$ which occurs when the estimated motion parameters match the actual motion parameters of the target. Furthermore, the focal quality indicator increases monotonically as the estimates deviate from the actual motion parameters. That characteristic can be exploited for efficient motion compensation.

The focal quality indicator will never satisfy the condition given at (8.38) because it has a limitation in the parameter measurement precision. Because the new focal quality indicator and the target motion estimation parameters are related by (8.35), the Cramer-Rao lower bound for the focal quality indicator can be defined as

$$\sigma_\gamma \geq (\sigma_v^2 + \sigma_a^2)^{1/2} \tag{8.40}$$

where

$$\sigma_v = \left[ \sum_{k=1}^{L} \frac{SNR\pi^2 T^2 \left( \sum_{j=1}^{\tilde{N}_k} |A_{kj}| \right)^2}{36 N^2 \Delta r^2 \left( \sum_{j=1}^{MN} |A_j| \right)^2 / (MN)^2} \right]^{-1/2} \tag{8.41}$$

and

$$\sigma_a = \left[ \sum_{k=1}^{L} \frac{SNR\pi^2 T^2 \left( \sum_{j=1}^{\tilde{N}_k} |A_{kj}| \right)^2}{2160 N^2 \Delta r^2 \left( \sum_{j=1}^{MN} |A_j| \right)^2 / (MN)^2} \right]^{-1/2} \quad (8.42)$$

Figure 8.1 shows Monte Carlo simulation results for the weighted least squares focal quality indicator. It is clear that the sample standard deviation of the focal quality indicator is in close agreement with its Cramer-Rao bound.

## 8.5 Motion Compensation Algorithm Via Weighted Least Squares

Based on our discussion, we propose the following motion compensation algorithm based on complex analysis and weighted least squares.

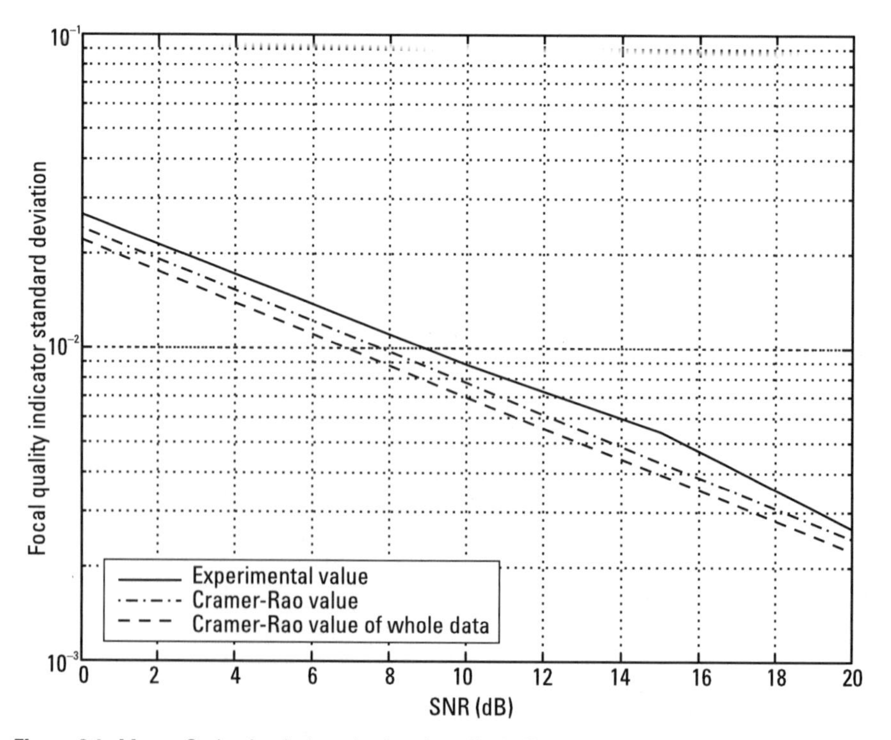


**Figure 8.1** Monte Carlo simulations for focal quality indicator standard deviation.

1. Reduce the noise of the echo transfer function via CFAR filtering.
2. Determine the phase standard deviation of the filtered echo transfer function.
3. Compensate the phase of the echo transfer function using rough estimates of the motion parameters.
4. Unwrap the phase of the compensated echo transfer function.
5. Perform complex analysis processing to estimate the instantaneous range of the target.
6. Estimate the motion parameters using weighted least squares.
7. Compute the focal quality indicator and the Cramer-Rao bound to assess the accuracy of the parameter estimates. If the focal quality indicator is lower than the Cramer-Rao bound for the focal quality indicator, go to step 10.
8. Update the motion parameter estimates and compensate the phase of the echo transfer function using the results from step 6.
9. Repeat steps 4 through 8 until the focal quality indicator value reaches a value lower than the Cramer-Rao bound of the focal quality indicator.
10. Generate a focused image of the target.

The flowchart in Figure 8.2 summarizes the procedure.

## 8.6 Simulation of Point Scatterers

In this section, a point target is generated and compensated using the proposed weighted least squares algorithm. For completeness, the performance of the algorithm is compared to that of the minimum entropy method.

### 8.6.1 Application to a Point Target

A two-point target was generated using the target parameters given in Table 8.1. We implemented an exhaustive search procedure over a two-dimensional $(v, a)$ search space confined to the region $v(-25, -15)$ m/s and $a(0, 10)$ m/s$^2$ to show the behavior of the weighted least squares focal quality indicator. In that demonstration, we assigned an initial velocity of $-20$ m/s and a constant acceleration of 5 m/s$^2$ to the target so that the actual motion parameters would correspond to the center of the selected search space. We evaluated the focal quality indicator over a total of $31 \times 31$ points in the search space corresponding to

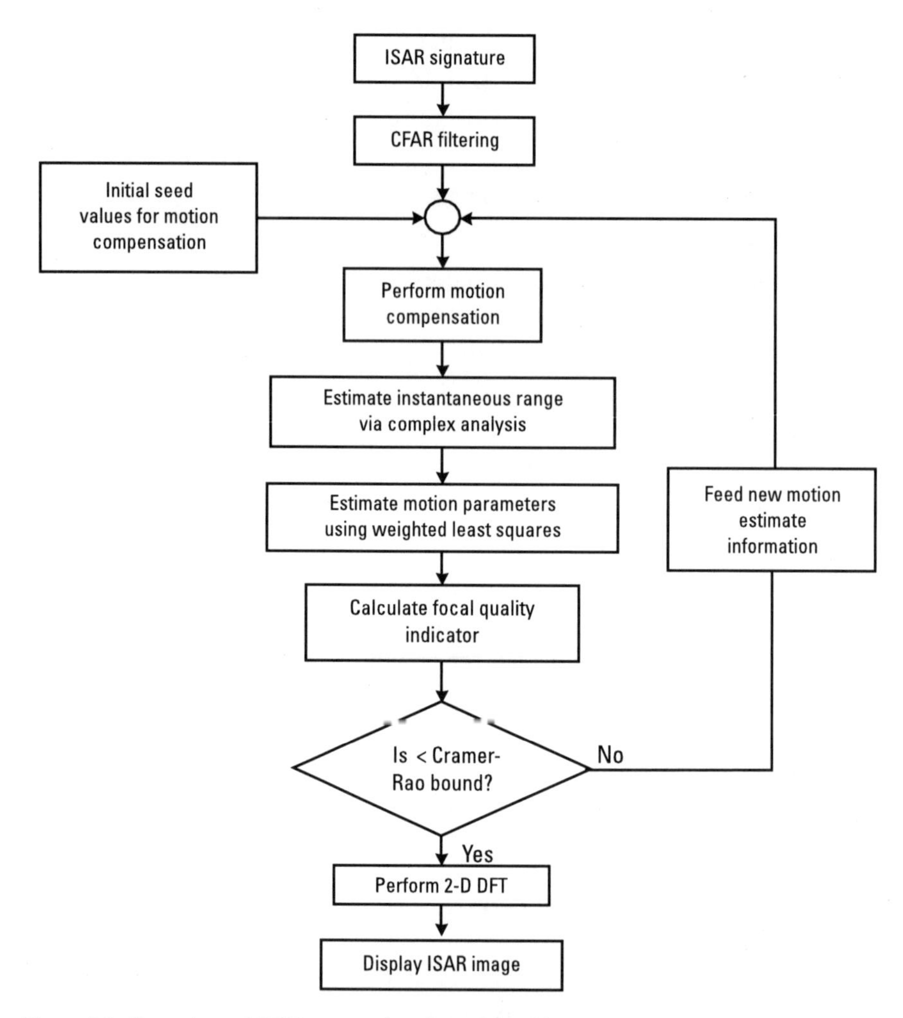


**Figure 8.2** Flow chart of ISAR processing via weighted least squares complex analysis.

**Table 8.1**
Sample Target Parameters

| Parameter | Symbol | Value |
|---|---|---|
| Target position | $r_0$ | −11m, −9m along LOS |
| Velocity | $v_0$ | −20 m/s |
| Acceleration | $a_0$ | 5 m/s$^2$ |
| Angular rotation rate | $\omega_0$ | 0 rad/s |
| SNR | SNR | 0, 5, 10, 15, 20 dB |

30 equal steps in velocity and acceleration. As shown in Figure 8.3, the focal quality indicator reaches a minimum at the center of the search space. Thus, it yields accurate motion parameter estimates.

Next, the proposed weighted least squares algorithm was applied to the signature of two point scatterers. The seed values for the target motion estimates were set to 0 m/s and 0 m/s$^2$. Table 8.2 shows the results of the motion estimation procedure from one iteration to the next. FQI denotes the focal quality indicator value. In the simulation, the SNR was set to 0 dB, and the resulting Cramer-Rao focal quality indicator bound was calculated to be 0.0250.

Table 8.2 shows that the algorithm converges so fast that the focal quality indicator reaches the Cramer-Rao bound after two iterations. A third iteration was included intentionally to show that in trying to achieve an accuracy greater than the Cramer-Rao bound, the velocity and acceleration estimates did not improve.

Table 8.3 shows simulation results for different SNRs. Here, the SNR ranges from 0 dB to 20 dB. In every case, the target motion estimates are extremely close to the actual motion parameters. In addition, the number of iterations needed in each case is limited to two, ensuring that the algorithm can be implemented in real time.

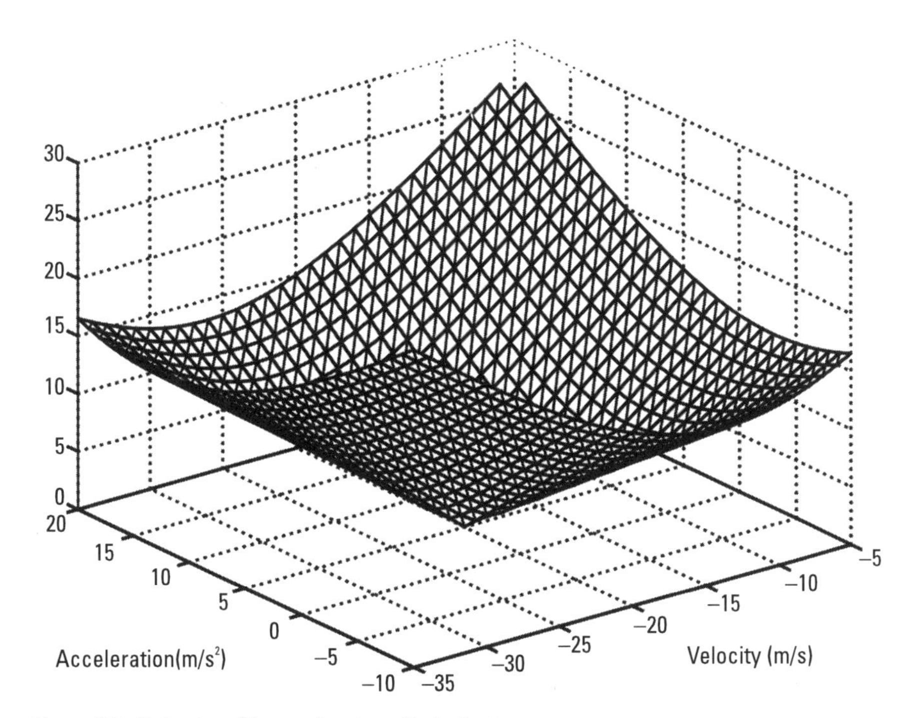


**Figure 8.3** Behavior of image focal quality indicator.

**Table 8.2**
Monte Carlo Simulation Results of Two Point Scatterers for SNR = 0 dB. The Cramer-Rao Bound for the Focal Quality Indicator is 0.0250

| Iteration | Velocity | Acceleration | FQI |
|---|---|---|---|
| 0 | 0 | 0 | 18.0449 |
| 1 | −17.3395 | 4.9960 | 2.6653 |
| 3 | −20.0048 | 4.9920 | 0.0031 |
| 4 | −20.0052 | 4.9921 | 0.0029 |

For illustrative purposes, two ISAR images are generated using the sample target parameters given in Table 8.1. Notice, however, that the angular rotation rate is set to 0.02 rad/s and the SNR is set to 0 dB. Figure 8.4 shows the uncompensated image of the two point scatterers. When the target signature is motion-compensated with the motion parameter estimates given in Table 8.3, the two point scatterers appear to be well focused, as shown in Figure 8.5.

### 8.6.2 Comparison of Weighted Least Squares and Entropy Results

The results obtained in Section 8.6.1 were compared to results obtained with the minimum entropy method, a standard method in ISAR motion compensation processing. Even though the entropy indicator is known for having a global minimum, closer inspection of its behavior reveals surface granularity in the search space near the minimum [2]. For this reason, it is possible that the entropy method may not converge to a unique solution when an optimization method to reduce the search time is used.

Table 8.4 was generated using the simplex method for the optimization. The simplex search method requires an initial guess of the motion parameters

**Table 8.3**
Monte Carlo Simulation Results of Two Point Scatterers for Five Different SNRs

| SNR | Velocity | Acceleration | Tolerance | Iterations |
|---|---|---|---|---|
| 0 | −20.0048 | 4.9920 | 0.0250 | 2 |
| 5 | −20.0033 | 4.9983 | 0.0141 | 2 |
| 10 | −19.9979 | 4.9975 | 0.0081 | 2 |
| 15 | −19.9999 | 4.9994 | 0.0048 | 2 |
| 20 | −20.0043 | 5.0033 | 0.0030 | 2 |

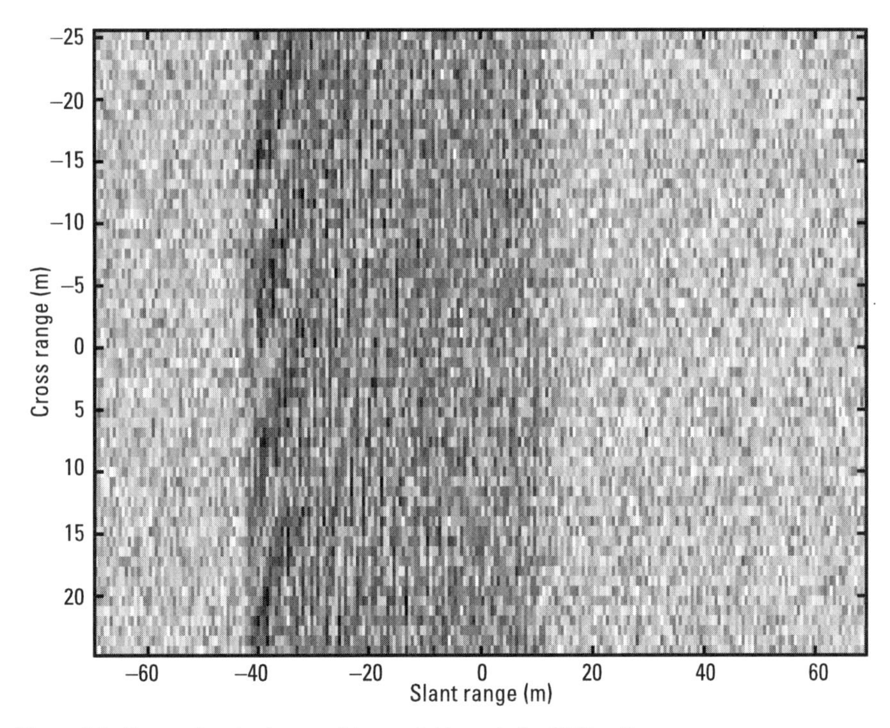


**Figure 8.4** Range-doppler image of two point targets for SNR = 0.

[3]. The initially guessed parameters are listed in the second and third columns of the table. The fourth and fifth columns summarize the results of the search. As shown here, when the initially guessed parameters were set to $v = 0$ m/s and $a = 0$ m/s$^2$, the entropy method failed to converge to the desired motion parameter values. On the other hand, when the initial guesses were sufficiently close to the actual motion parameters (as shown on the third row), the simplex method converged to the actual motion parameters closely.

## 8.7 ISAR Image Formation

This section examines the effectiveness in applying the weighted least squares algorithm to a real-world target signature. The chosen target ISAR signature was that of a Boeing 727 aircraft. Following the procedure in Figure 8.2, the resulting estimates were $v_0 = -9.434$ m/s and $a_0 = 4.60$ m/s$^2$. Figures 8.6, 8.7, and 8.8 show the image formation results when motion compensation is performed using those estimates but different windows.

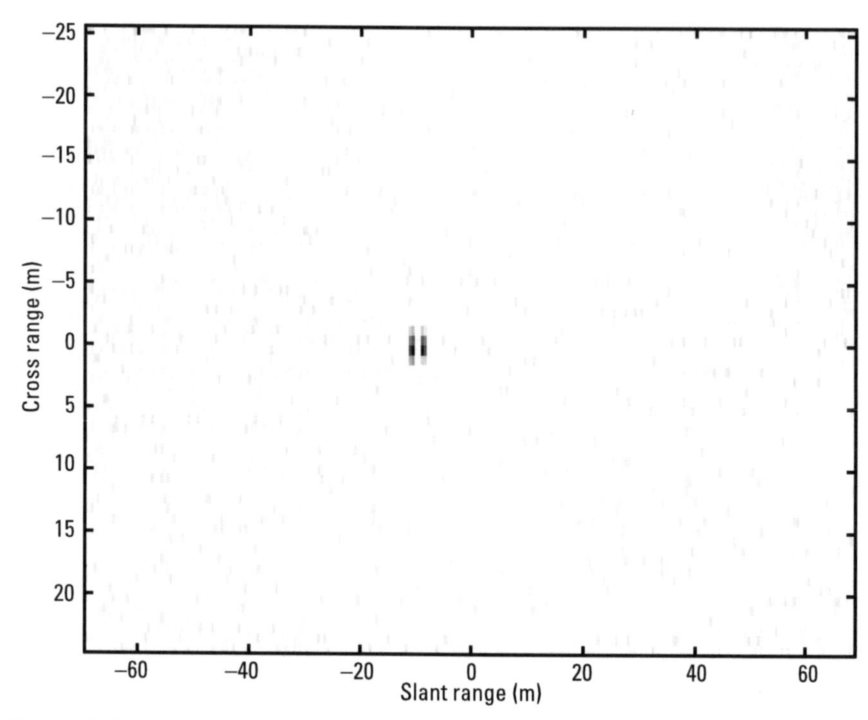


**Figure 8.5** Motion compensated range-doppler image of two point targets for SNR = 0.

**Table 8.4**
Simplex Search Results Using Exponential Entropy

| SNR | Guessed velocity | Guessed acceleration | Velocity | Acceleration | Entropy | Iterations |
|---|---|---|---|---|---|---|
| 0 | 0 | 0 | 0.0007 | −0.0035 | 2.7180 | 100 |
| 100 | 0 | 0 | 0.0036 | −0.0017 | 2.7178 | 100 |
| 0 | −18 | 7 | 20.8008 | 4.9994 | 2.7169 | 100 |
| 100 | −20 | 5 | −20 | 5 | 2.7180 | 3 |

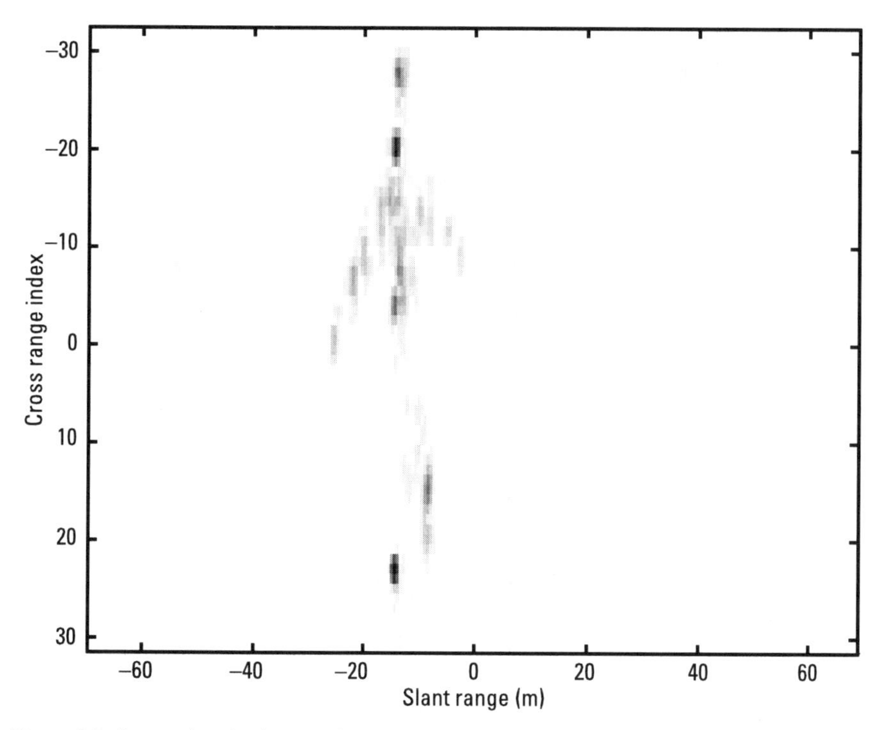


**Figure 8.6** Range-doppler image of motion-compensated Boeing 727 using Hann window.

Sidelobe artifacts occur in image reconstruction when a finite-extent Fourier transform is used. Conventional windows such as the Hann or Hamming weighting functions reduce these artifacts at the expense of reducing mainlobe resolution. Spatially variant apodization (SVA) as suggested in [4, 5] also reduce the sidelobe artifacts while preserving the mainlobe resolution.

Figure 8.6 uses the Hann window for sidelobe reduction. Although the sidelobes are suppressed significantly, there is loss of resolution. Figure 8.7 was generated by using a rectangular window. Even though the resolution is the highest possible, strong sidelobes corrupt the image. Figure 8.8 was generated using the SVA method. Even though the sidelobe is suppressed as much as in the case of Hann window, the resolution is almost the same as when the rectangular window is used. In Figure 8.8, the prominent scatterers of the Boeing 727 aircraft distributed in range and cross-range are clearly discernible.

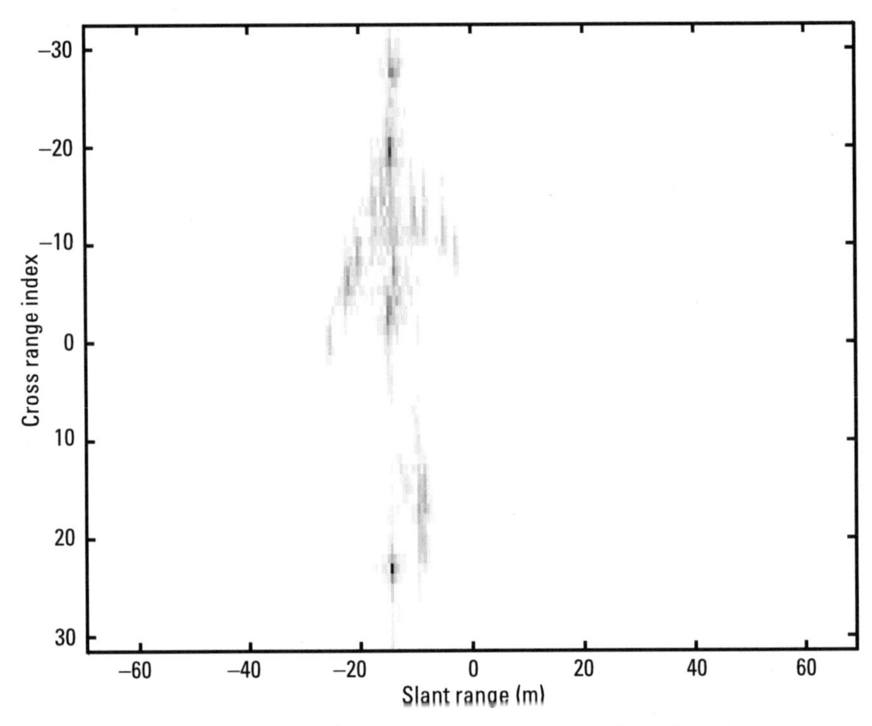


**Figure 8.7** Range-doppler image of motion compensated Boeing 727 using rectangular window.

## 8.8 Summary

The use of weighted least squares for motion estimation was justified. This chapter showed that the weighting matrix can be derived from the magnitude information of the ISAR signature. That weighting matrix gives more importance to a segment with large magnitude and great number of samples in a segment; the phase slope within this segment is highly accurate due to its high-SNR nature. Thus, the weighted least squares estimation was applied to segments of varying amplitude with different numbers of samples in a segment. In this chapter, the polynomial parameters obtained through the weighted least squares were used to define a focal quality indicator, and the Cramer-Rao bound for the focal quality indicator was derived. When the least squares method is combined with complex analysis to motion compensate simulated and actual target signatures, the method yielded extremely accurate motion estimates.

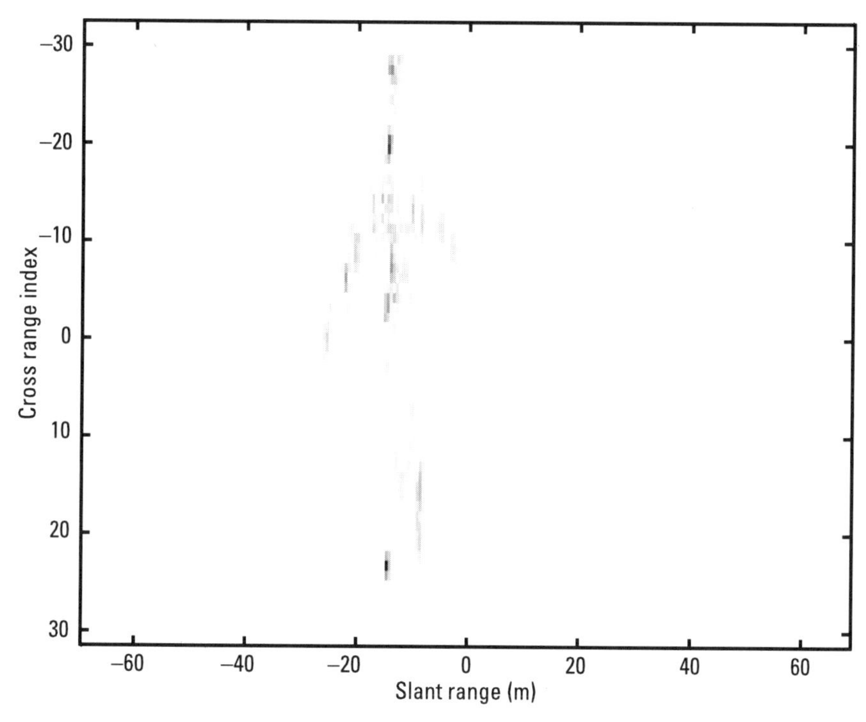


**Figure 8.8** Range-doppler image of motion compensated Boeing 727 using SVA apodization.

# References

[1] Hostetter, G. H., M. S. Santina, and P. D'Carpio-Montalvo, *Analytical, Numerical, and Computational Methods for Science and Engineering*, Englewood Cliffs, NJ: Prentice Hall, 1991.

[2] Wehner, D. R., *High-Resolution Radar*, Norwood, MA: Artech House, 1995.

[3] Press, W. H., B. P. Flannery, S. A. Teukolsky, and W. T. Vetterling, *Numerical Recipes: The Art of Scientific Computing*, New York: Cambridge University Press, 1986.

[4] Stankwitz, H. C., R. J. Aallaire, and J. R. Fineup, "Nonlinear Apodization for Sidelobe Control in SAR Imagery," *IEEE Trans. on Aerospace and Electronic Systems*, Vol. 31, No. 1, Jan. 1995.

[5] Lee, A. C., and D. C. Munson, Jr., "Effectiveness of Spatially-Variant Apodization," *Proc. IEEE Internat. Conf. on Image Processing*, Washington, DC, Oct. 1995, pp. 147–150.

# 9

# Signal Analysis and Synthesis Using Short-Time Fourier Transform

To accomplish selective motion compensation, target signatures must be discriminated and processed separately. The filtering process is performed in the time-frequency domain. The first section of this chapter discusses the analysis and synthesis algorithms used to separate radar signals with non-overlapping time-frequency representations.

## 9.1 The Continuous Short-Time Fourier Transform

The decomposition of superimposed target signatures can be performed via the short-time Fourier transform (STFT), a two-dimensional linear transformation that conveys information in time and frequency via a sliding analysis window:

$$STFT_X(t, f) = \int_{t'} [x(t')w^*(t' - t)]e^{-j2\pi ft'}dt' \tag{9.1}$$

The STFT involves the Fourier transform of the windowed signal at time $t$, as indicated in Figure 9.1.

This two-dimensional representation has more signal structure information than the one obtained from the Fourier transform alone. Figure 9.2 shows the multicomponent signal $x(t) = x_1(t) + x_2(t) + x_3(t) + x_4(t)$ and its Fourier spectrum. As it can be seen from the time domain information in Figure 9.2(a), the signal contains a pulse component at 1.85 sec, which is the only informa-

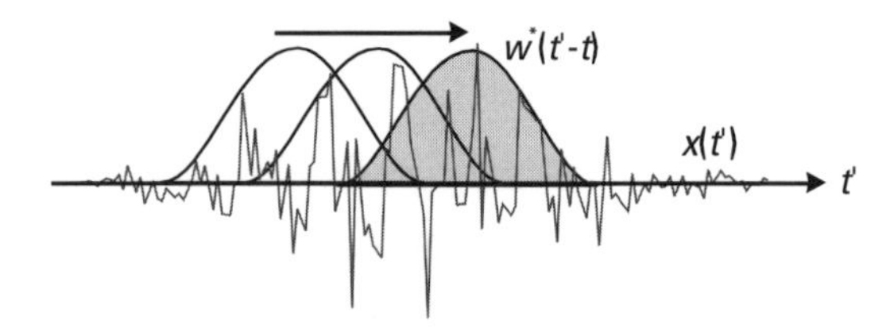


**Figure 9.1** The STFT of a signal is obtained by computing the Fourier transform of a block while sliding a time analysis window *w*.

tion that can be inferred with certainty from the figure. From Figure 9.2(b), which shows the Fourier transformation, the spectrum shows a sinusoidal component at 100 Hz. No more information can be extracted from the signal itself or the spectrum.

When the STFT is used, the time and frequency information offered by that representation shows how the signal frequency of each component evolves

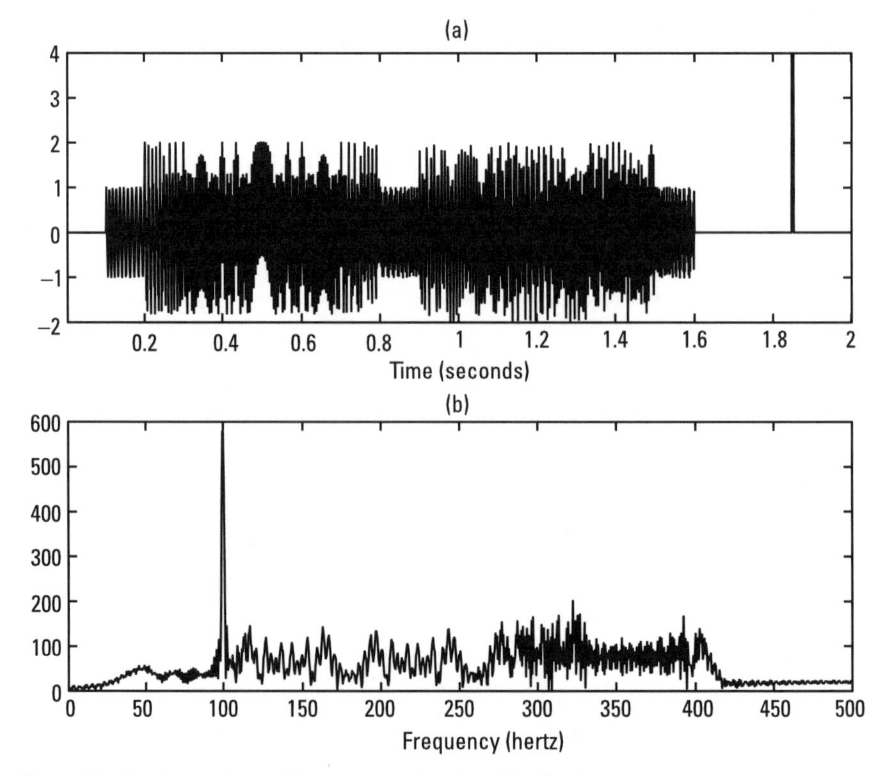


**Figure 9.2** Real part of a multicomponent signal and its Fourier spectrum.

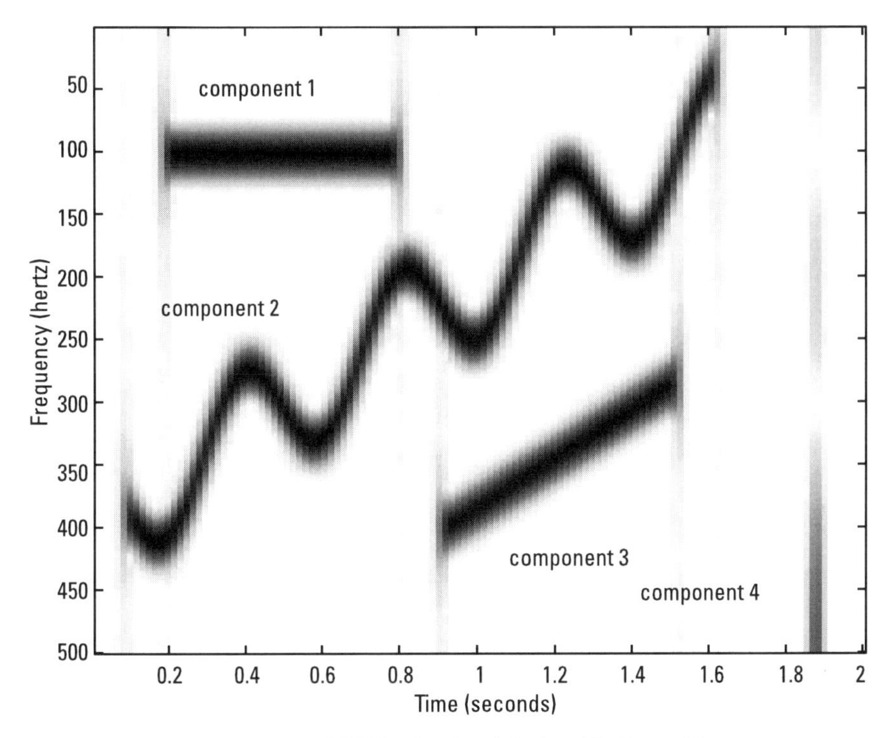


**Figure 9.3** Absolute value of the STFT for the signal depicted in Figure 9.2.

with time, as shown in Figure 9.3. If components 1, 2, and 3 are complex exponential signals with a time-varying phase, the instantaneous frequency of each of these components can be expressed as

$$f_I(t) = \frac{1}{2\pi}\frac{d\phi(t)}{dt} \tag{9.2}$$

For component 1, it can be seen that (9.2) is constant, that is,

$$\frac{1}{2\pi}\frac{d\phi_{c1}(t)}{dt} = f_{c1} \tag{9.3}$$

Thus the component can be expressed as

$$x_1(t) = A\exp[j\phi_{c1}(t)] = A\exp[j2\pi f_{c1}t] \quad \text{for } 0.2 < t < 0.8 \tag{9.4}$$

where we can extract the time duration from its TF representation by looking at the time axis; its frequency $f_{c1} = 100$ Hz is shown in the frequency axis.

For the chirp signal that forms component 3, an expression can be given by

$$x_3(t) = B\exp[j2\pi(f't + f''t^2)] \quad \text{for } 0.9 < t < 1.5 \tag{9.5}$$

Using (9.3), the instantaneous frequency of component 3 is

$$f_I(t) = \frac{1}{2\pi}\frac{d\phi_{c3}(t)}{dt} = f' + 2f''t \tag{9.6}$$

The TF plane shows a negative slope (with time, frequency decreases) of approximately 200. Thus, $2f'' = -200$ and therefore $f' = -100$. At $t = 0.9$, the frequency of this component is approximately 380 Hz. Using the instantaneous frequency given in (9.6), we have $f' - 2(100)(0.9) = 380$, and $f' =$ 560 Hz.

For a phase modulation such as the one for component 2, it can be inferred from its TF structure that the component has the following form:

$$x_2(t) = C\exp[j2\pi(f't + f''t^2 - D\sin(2\pi f'''t)] \quad \text{for } 0.1 < t < 1.6 \tag{9.7}$$

where

$$f_I(t) = \frac{1}{2\pi}\frac{d\phi_{c2}(t)}{dt} = f' + 2f''t - 2\pi D\cos(2\pi f'''t) \tag{9.8}$$

The parameter $f''$ is approximately equal to $-100$, because it has the same slope as component 3. It can be seen that the oscillatory period of that component is around 0.4 sec. Thus, $f''' = 2.5$ Hz. The parameters $f'$ and $D$ can be found by evaluating two points in (9.8). From the TF representation, it can be seen that at $t = 1.6$ sec, (9.8) is approximately 50 Hz. And at $t = 1$, (9.8) is 250 Hz. Solving for the last unknown parameters yields the values $f' = 235$ and $D = 5.57$.

For the component amplitudes, since the same intensity is shown in the TF plane for all of them but the transitory pulse, we can conclude that $A = B = C$. The pulse amplitude can be found from Figure 9.2(a).

From that example, parameters such as target velocity and acceleration might be estimated from a TF representation. However, the estimation of those parameters, as well as the estimation of the parameters in the example, are only estimations which usually result in an unsatisfactory kinematic parameter estimation for ISAR motion compensation purposes.

The STFT shares many of the properties of the Fourier transform. Some common properties are the following [1]:

- The STFT is a linear transformation:

$$STFT_{X+Y}(t, f) = STFT_X(t, f) + STFT_Y(t, f) \tag{9.9}$$

- The STFT preserves time and frequency shifts:

$$\tilde{x}(t) = x(t)e^{j2\pi f_0 t} \Rightarrow STFT_{\tilde{x}}(t, f) = STFT_x(t, f - f_0) \tag{9.10}$$

and

$$\tilde{x}(t) = x(t - t_0) \Rightarrow STFT_{\tilde{x}}(t, f) = STFT_x(t - t_0, f) \tag{9.11}$$

- The STFT transform can also be expressed in terms of the signal and window Fourier spectra:

$$STFT_X(t, f) = e^{j2\pi tf} \int_{t'} X(f')\Gamma^*(f' - f)e^{j2\pi f't}df' \tag{9.12}$$

where $\Gamma(f)$ denotes the Fourier transform of the analysis window.

The choice of the analysis window affects time and frequency resolution. Besides the shape characteristics of the window, the length affects the time duration $\Delta T$ and bandwidth $\Delta\beta$, which yields the uncertainty product [2]:

$$\Delta\beta\Delta T \geq \frac{1}{4\pi} \tag{9.13}$$

Thus, the choice of a better time resolution reduces frequency resolution and vice versa, as shown in Figures 9.4 and 9.5. The signal in both figures consists of a concatenation of sinusoidal segments. Frequency resolution is better using a longer analysis window, as depicted in Figure 9.4, than using a shorter

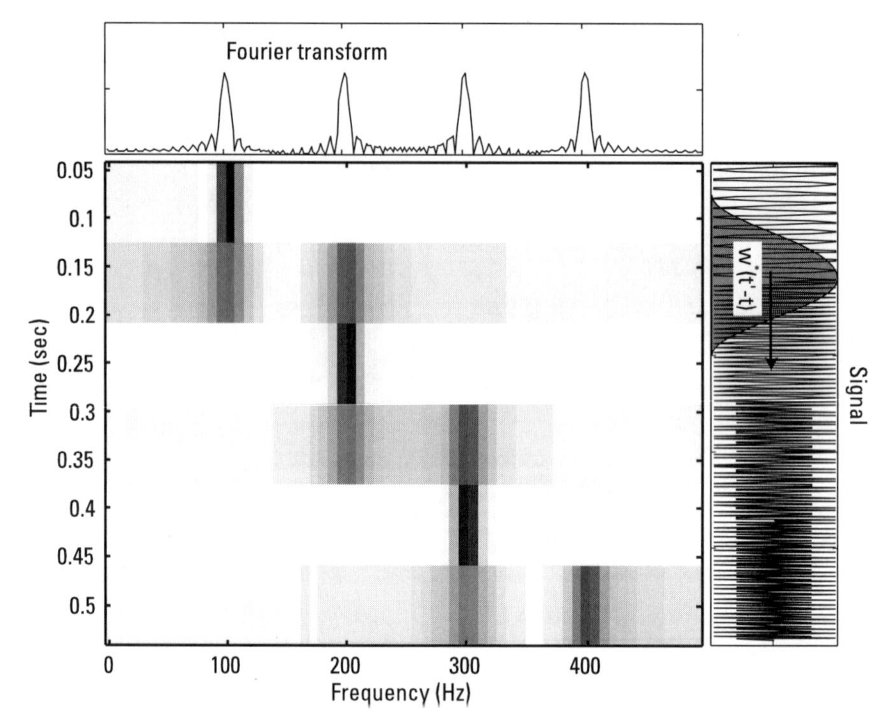


**Figure 9.4** STFT of simulated signal using a long duration analysis window.

window, like that used in Figure 9.5. Time resolution degrades in the other direction. Thus, the beginning and ending times of each sinusoidal segment are easier to detect using shorter windows, as depicted in Figure 9.5.

The squared absolute value of the STFT belongs to the quadratic time-frequency type of representations. This particular quadratic representation is known as the spectrogram, denoted as

$$SG_X(t, f) = \left| \int_\tau [x(\tau) w^*(\tau - t)] e^{-j2\pi f \tau} d\tau \right|^2 \tag{9.14}$$

In general, the quadratic representations $T_X(t, f)$ seek to satisfy properties such as time and frequency marginals, denoted as

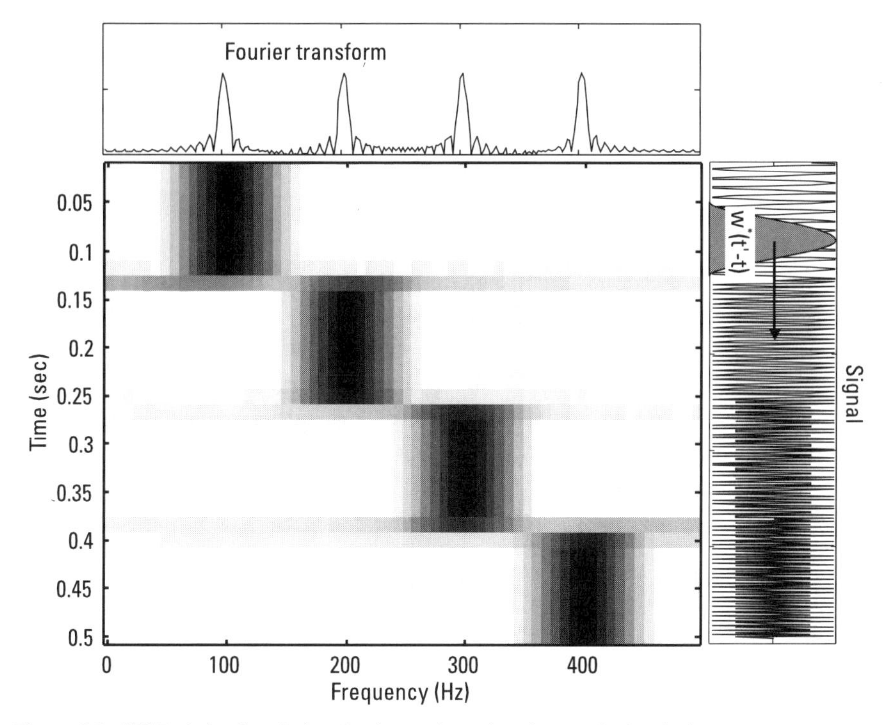


**Figure 9.5** STFT of simulated signal using a short duration analysis window.

$$\int_f T_X(t, f)df = |X(f)|^2 \tag{9.15}$$

and

$$\int_t T_X(t, f)dt = |x(t)|^2 \tag{9.16}$$

respectively.

Those representations are compared to density functions encountered in probability. Another desirable property is that every outcome is always real-valued and positive so that

$$T_X^*(t, f) = T_X(t, f) = |T_X(t, f)| \tag{9.17}$$

All quadratic representations, except the STFT and the Gabor, satisfy the following superposition principle [1]:

$$x(t) = c_1 x_1(t) + c_2 x_2(t) \Rightarrow T_X(t, f) = |c_1|^2 T_{x_1}(t, f) + |c_2|^2 T_{x_2}(t, f) + c_1 c_2^* T_{x_1,x_2}(t, f) + c_2 c_1^* T_{x_1,x_2}(t, f) \quad (9.18)$$

where $T_X(t, f)$ are the self-terms of the signal and $T_{x_1,x_2}(t, f)$ are the cross-terms.

As stated in [3], the main advantage of the STFT over quadratic TF representations is that no cross-terms appear on the TF plane. That characteristic of the STFT, and thus the spectrogram, comes from the fact that the representation operates on the signal $x(t)$ rather than a bilinear function, as in the case of other quadratic distributions such as the Wigner-Ville represented by

$$WV_X(t, f) = \int_\tau \left[ x\left(t - \frac{1}{2}\tau\right) x^*\left(t + \frac{1}{2}\tau\right) \right] e^{-j2\pi f\tau} d\tau \quad (9.19)$$

Cross-terms in the TF representation of a target signature can make the detection difficult since they can overlap with different target signature components. Cross-term intensities in the TF plane can also be higher than the self-term intensities of interest, complicating the detection process. Those effects are shown in Figure 9.6 and result from the Wigner-Ville representation, which shows the cross-term response between the two modulated Gaussian components of the analysis signal. In contrast, the spectrogram shown in Figure 9.7 lacks cross-terms associated with quadratic distributions.

Note also that the intensity of the two modulated Gaussians do not overlap in the time-frequency plane. That is not the case using the Fourier transform as shown in the right plots of Figures 9.6 and 9.7. If component decomposition is desirable, the nonoverlapping of the time-frequency signatures becomes important in the design of an effective technique for multicomponent signal decomposition and reconstruction.

There is a class of time-frequency representations, $\rho(t, f)$, that minimizes the cross-terms. That class, known as Cohen's class [4], is represented by the family of time-frequency representations obtained by changing the kernel $g(v, \tau)$ in

$$\rho(t, f) = \iint M(v, \tau) e^{-j2\pi vt - j2\pi\tau f} dv d\tau \quad (9.20)$$

where

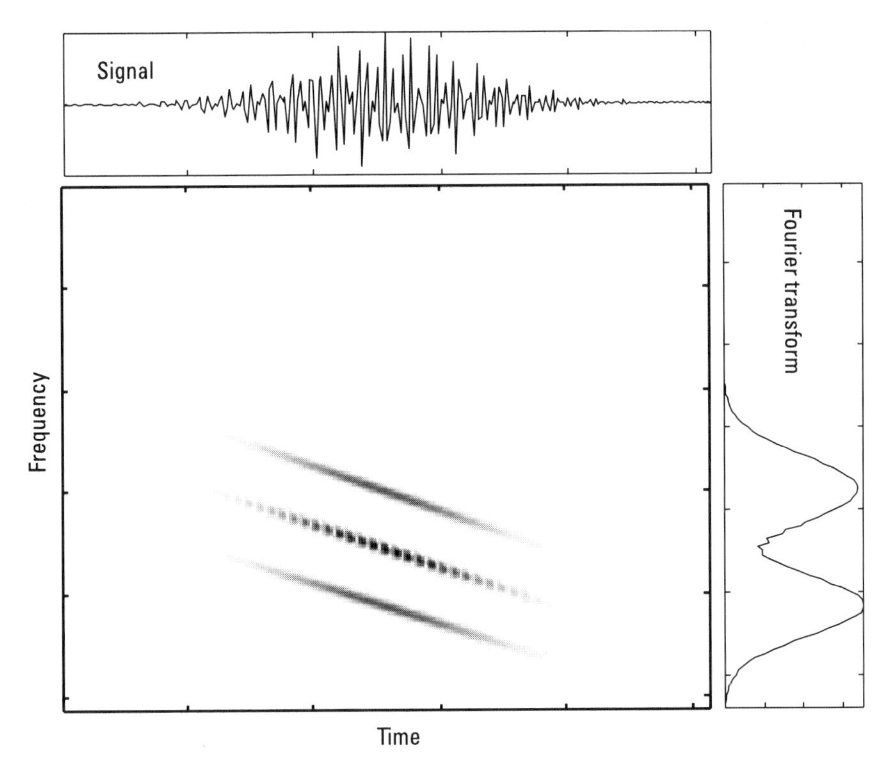


**Figure 9.6** Example of Wigner-Ville signal representation.

$$M(v, \tau) = g(v, \tau) \int_{-\infty}^{\infty} x^*\left(u - \frac{1}{2}\tau\right) x\left(u + \frac{1}{2}\tau\right) e^{j2\pi v u}\, du \qquad (9.21)$$

The kernel is thus a two-dimensional filter that eliminates the cross-terms. For the Wigner-Ville representation, the kernel is equal to unity; that is, no cross-term suppression is used.

Some of the different kernels proposed form the following representations:

- The Margenau-Hill representation is formed with the use of a kernel expressed as

$$g(v, \tau) = \cos(\pi v \tau) \qquad (9.22)$$

- The Choi-Williams representation is formed with its kernel defined as

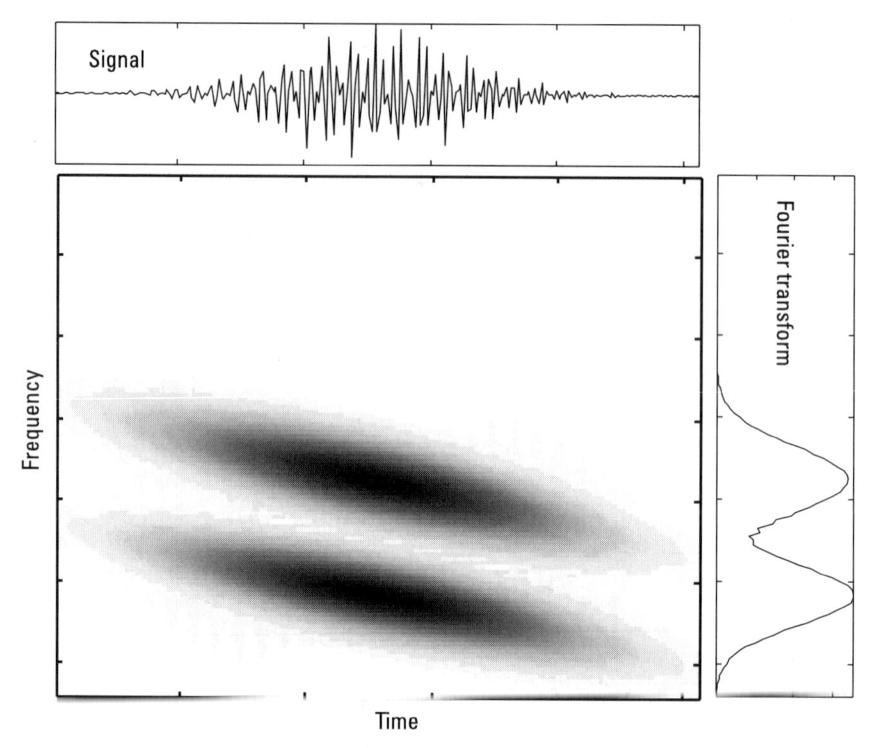


**Figure 9.7** Example of STFT signal representation with no cross-terms between components.

$$g(v, \tau) = e^{-v^2\tau^2/\sigma} \tag{9.23}$$

where $\sigma$ defines the volume of the filter.

- The Zhao-Atlas-Marks representation, using a cone-shaped kernel as proposed in [5], considers the kernel Fourier transform in $v$ to have the following form

$$G(t, \tau) = \begin{cases} 1 & \text{for } -\dfrac{|\tau|}{2} \leq t \leq \dfrac{|\tau|}{2} \\ 0 & \text{else} \end{cases} \tag{9.24}$$

Figures 9.8, 9.9, and 9.10 show the resulting time-frequency representations obtained when those kernels are used for the same signal as that in Figures 9.4 and 9.5.

Although the Choi-Williams and Zhao-Atlas-Marks TF representations seem to have good resolution without cross-terms, eliminating those cross-

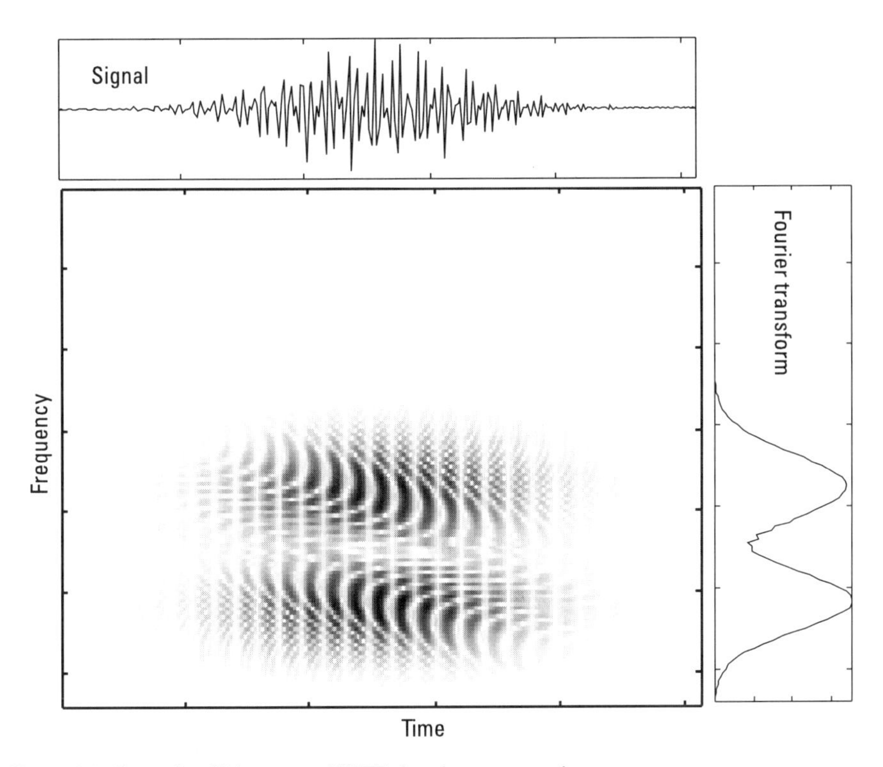


**Figure 9.8** Example of Margenau-Hill TF signal representation.

terms does reduce energy that belongs to the self-terms. Thus, signal reconstruction may not be feasible in all cases.

## 9.2 The Discrete STFT

To implement the STFT in the form of a computer algorithm, it is necessary to take equidistant samples from the time-frequency plane. Denoting $T_s$ and $F_s$ as the sampling time and frequency periods, time and frequency samples, $t_n$ and $f_k$, are taken from the expression in (9.1)

$$STFT_X(nT_s, kF_s) = \int_{t'} [x(t')w^*(t' - nT_s)]e^{-j2\pi(kF_s)t'} \, dt' \quad (9.25)$$

which, for discrete time and frequency, takes the form

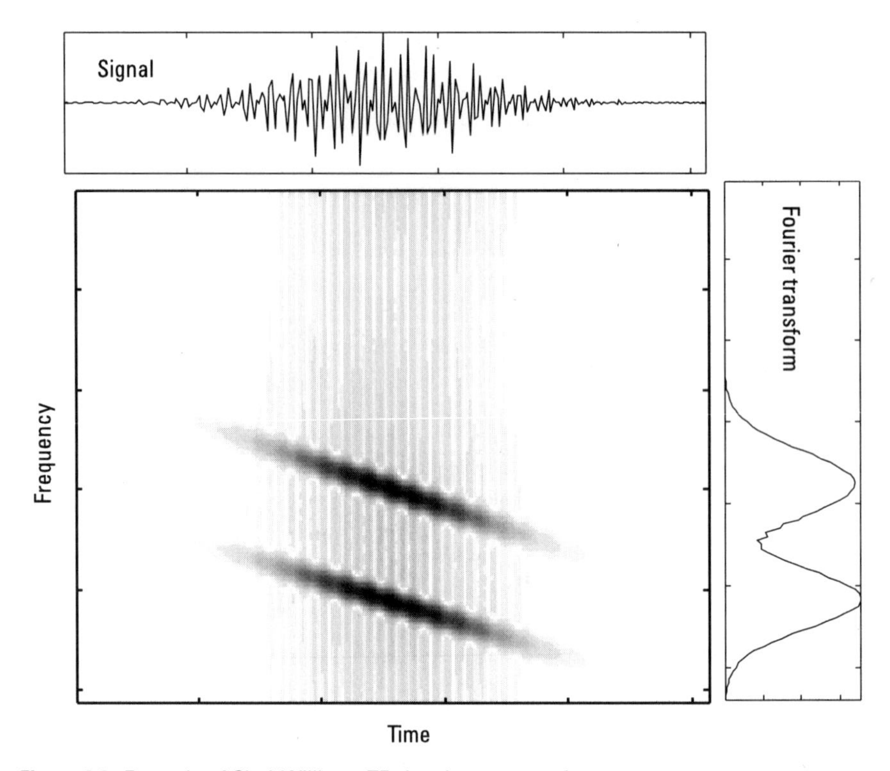


**Figure 9.9** Example of Choi-Williams TF signal representation.

$$STFT(t_n, f_k) = \sum_{l = t_n D - \frac{T}{2}}^{t_n D + \frac{T}{2} - 1} w(t_n D - l)x(l)e^{-j2\pi fkl/T} = \Im[w(t_n D - l)] \tag{9.26}$$

where $x(l)$ represents the sampled signal, $w(t_n D - l)$ is a time domain window whose location is a multiple of $D$ samples in time, and $\Im$ corresponds to the evaluation of $T$ uniformly spaced samples via the discrete-time Fourier transform. Thus, at each time index $t_n$, a discrete Fourier transform is computed on a block of samples weighted by a window function that is centered at location $t_n D$.

When implemented via the fast Fourier transform (FFT), the STFT method is highly computationally efficient. Likewise, its inverse transformation is also very effective and simple to implement.

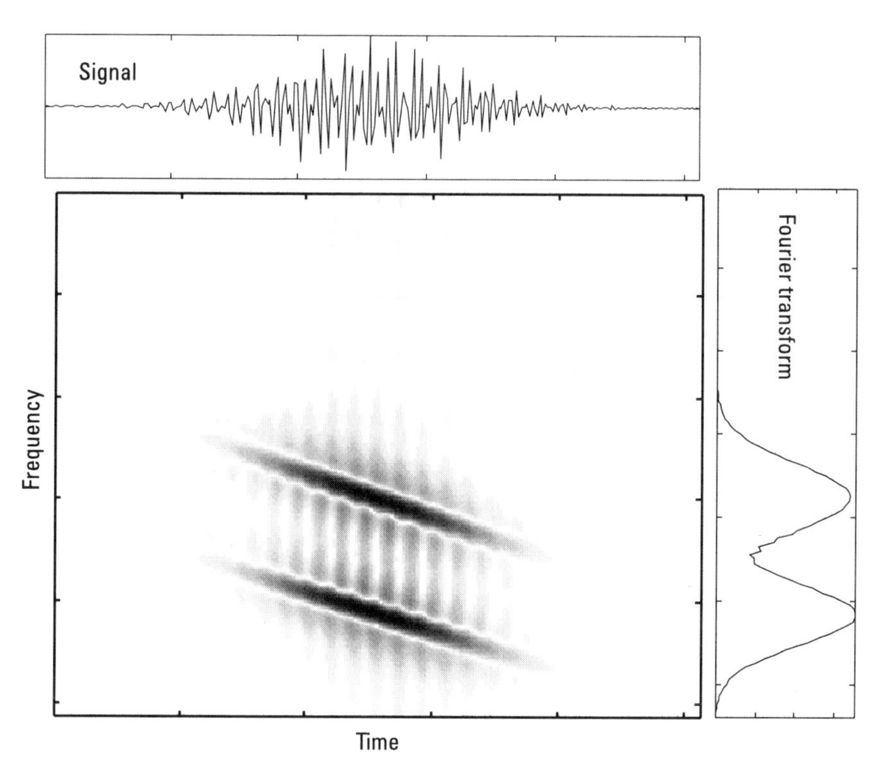


**Figure 9.10** Example of Zhao-Atlas-Marks TF signal representation.

## 9.3 Parallel Processing of the STFT

The use of data blocks during the computation of a discrete STFT can be exploited to use multiple processors. Parallel processing can be done if the number of processors required to process $L$ number of blocks is available. The group of processors working in tandem can be viewed as an analysis window of length $LM$, where $M$ represents the number of samples for each block, that is, the length of an analysis window in the STFT.

Assume that two processors ($L = 2$), $P_1$ and $P_2$, are available for computing the STFT of a signal like the one shown in the upper plot of Figure 9.11. Two options are possible. First, for each pair of blocks, the two processors can intercalate the work for each consecutive block. That is, the blocks obtained using the windows $w(l)$, $w(2D - l)$, and $w(4D - l)$ are processed by $P_1$. The blocks obtained using the windows $w(D - l)$ and $w(3D - l)$ are processed by

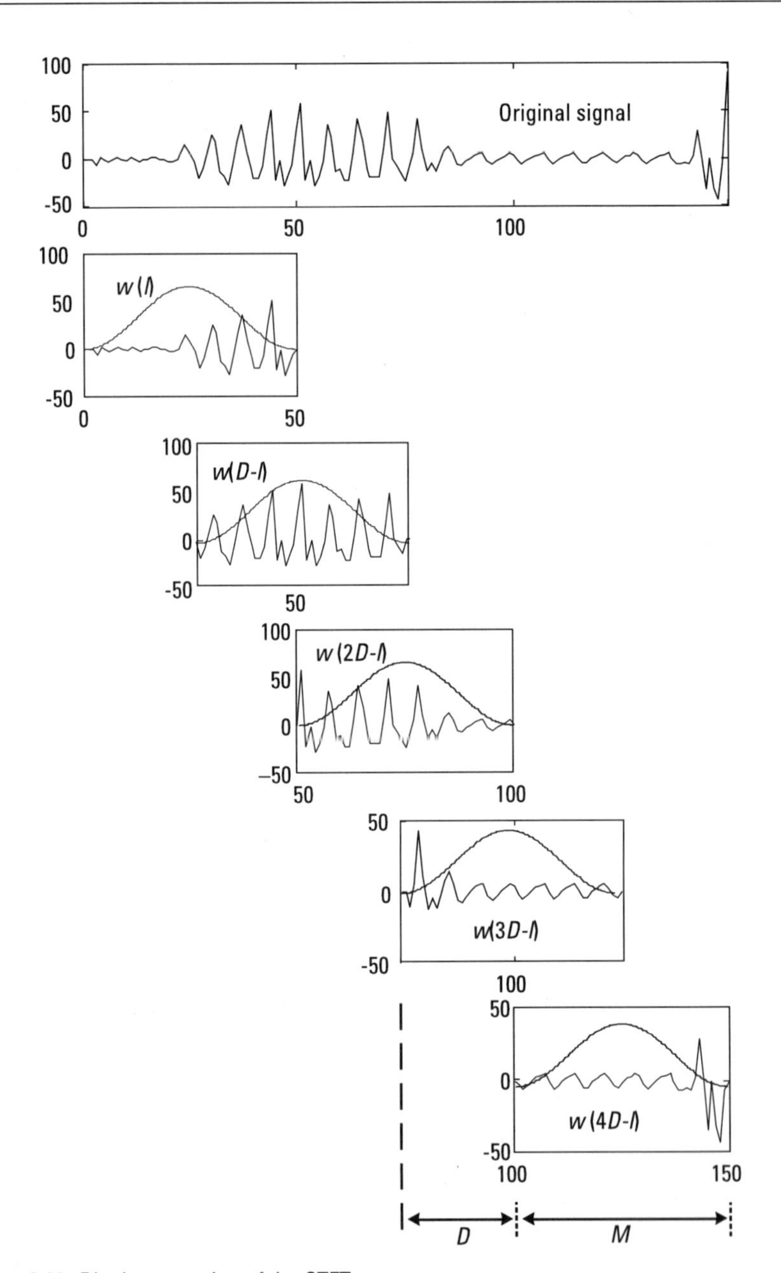


**Figure 9.11** Block processing of the STFT.

$P_2$. Second, another processor job distribution can be selected so that $w(l)$, $w(D - l)$, and $w(2D - l)$ are processed using $P_1$, and $w(3D - l)$ and $w(4D - l)$ by $P_2$.

In both cases, each processor added to the computation of the STFT represents almost a 100% increase in computational time. The work overhead produced by assigning and tracking the block position for each computation is negligible compared to the computational time required for a single Fourier transform.

Parallel processing of the STFT is possible since the window multiplication and discrete Fourier transform operations required for each processor are independent of other processor results.

## 9.4 STFT Synthesis Using the Overlap-Add Method

Prior to selective motion compensation, a time-varying filter must be applied to the time-frequency domain signature. Subsequently, the time-frequency representation of each signature must be reconstructed in the original domain. To compensate motion and obtain focused images, the minimum entropy is computed in the frequency domain of the signatures. Next, an approach for reconstructing a time-frequency representation in the context of STFT processing is presented.

### 9.4.1 Overlap-Add Method for Inverse STFT Calculation

The procedure for performing the synthesis from the STFT is the overlap-add (OLA) method [6]. According to the OLA procedure, the synthesis is based on the following condition for an analysis window:

$$\sum_{n=-\infty}^{\infty} w(t_n D - l) = 1 \tag{9.27}$$

The reconstruction can be achieved perfectly under that condition by multiplying both sides of (9.27) by the signal $x(l)$, which yields

$$\sum_{n=-\infty}^{\infty} w(t_n D - l)x(l) = x(l) \tag{9.28}$$

That summation is the inverse Fourier transform of (9.26) for a fixed index $n$, that is

$$x_p(l) = \Im^{-1}\{STFT_x(t_n, f_k)\} \tag{9.29}$$

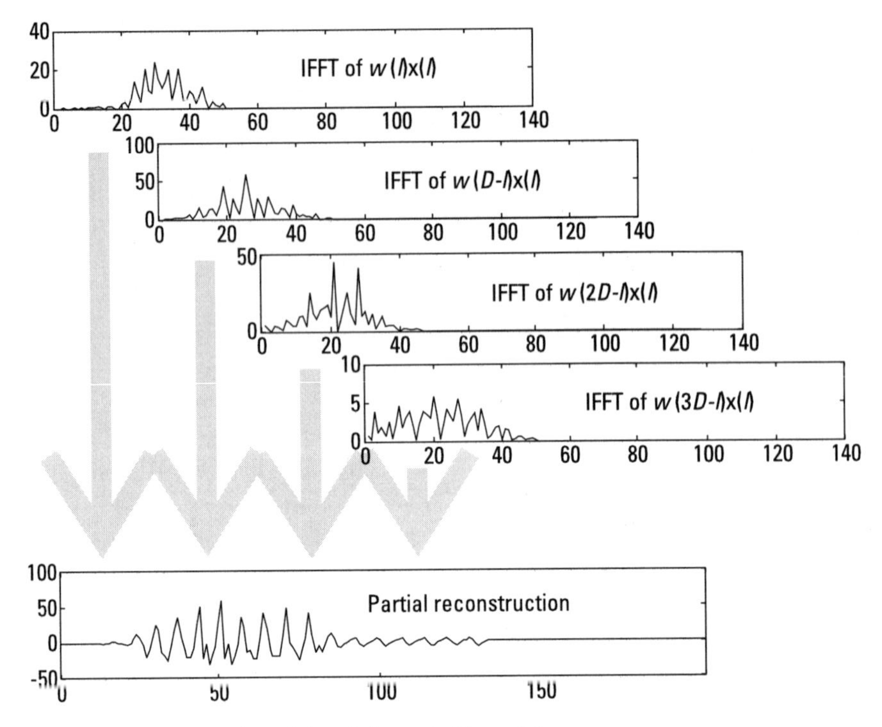


**Figure 9.12** Reconstruction of signal shown in Figure 9.9.

where $x_p(l)$ represents a partial reconstruction. Therefore, the reconstruction procedure is performed by calculating an inverse FFT (IFFT) to obtain a windowed block of $L$ samples, shifting the result by $D$ samples in time and adding it to the partial reconstruction. Figure 9.12 shows this process using the example in Figure 9.11.

### 9.4.2 Overlap Value for Optimum Reconstruction Using the Inverse STFT

In (9.27), the parameter $D$ should be chosen so that the summation of the window-shifted replicas has a constant value. A procedure is presented in [6] to find the conditions for perfect reconstruction by considering the STFT calculation as a filter bank decomposition with an analysis carried out in the frequency domain. This section derives an alternative solution for perfect reconstruction using OLA in the time domain.

Equation (9.27) can be interpreted as a convolution of a window with an impulse train:

$$\sum_{n=-\infty}^{\infty} w(t_n D - l)x(l) = w(l) * S_D(l) \tag{9.30}$$

where $w(-l) = w(l)$ is assumed, and

$$S_D(l) = \sum_{n=-\infty}^{\infty} \delta(t_n D - l) \tag{9.31}$$

The Fourier transform of (9.30) is

$$W(f)\left[\frac{2\pi}{D}\sum_{m=-\infty}^{\infty} \delta\left(f - \frac{2m\pi}{D}\right)\right] = 2\pi\delta(f) \tag{9.32}$$

Thus, for (9.32) to be valid, the zeros of $W(f)$ must lie at positions that are multiples of $f = 2\pi/D$, except at $f = 0$ where (9.30) becomes $W(0)/D = 1$, as indicated in Figure 9.13. Therefore, the summation of (9.27) is equal to the constant $W(0)/D$. That constant must be multiplied by the partial reconstruction $\hat{x}(l)$ obtained from the summation of shifted inverses. The final expression takes the form

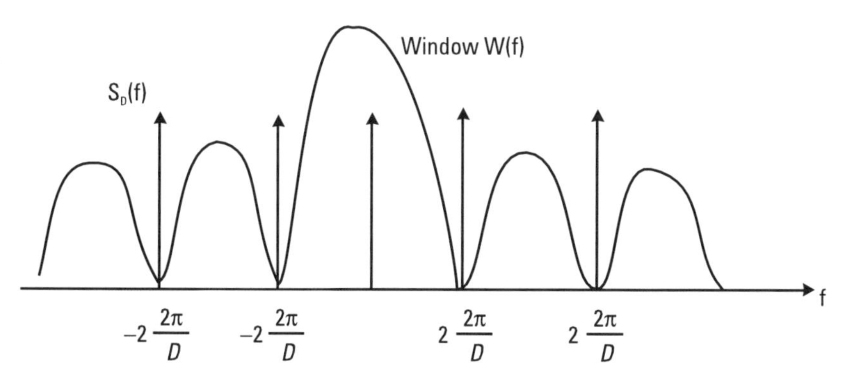


**Figure 9.13** Impulse positions must be chosen so they match the position of the zeros in the frequency domain.

$$x(l) = \frac{D}{\sum_{n=0}^{L-1} w(t_n)} x_p(l) \tag{9.33}$$

Perfect reconstruction clearly depends on the type of window used. The only restriction on the window is that its frequency representation must show periodic zeros. The shape of the window spectrum will define the value of $D$ and the multiplicative constant $W(0)/D$. It should be noted that the condition in (9.32) for perfect reconstruction can be approximately satisfied by having a band-limited window with bandwidth $\beta_o$. Although every value of $D < 1/\beta_o$ will work, this cannot be satisfied entirely for a finite-duration window, but it is practically true if the window has very small sidelobes. In that case, half the mainlobe width of the window determines its essential bandwidth. An example of how to choose the parameter $D$ during reconstruction is shown next.

### 9.4.3 Reconstruction Using the Hamming Window

Let us define a discrete time domain Hamming window as [7]:

$$w(t_n) = 0.54 + 0.46 \cos\left(\frac{2\pi}{N} t_n\right) \quad \text{for } t_n = \frac{-N}{2}, \ldots, -1, 0, 1, \ldots, \frac{N}{2} \tag{9.34}$$

The shifting value $D$ must be computed such that (9.27) is true for this particular window. The discrete Fourier transform of (9.34) is given by

$$W(f) = 0.54 \exp\left[-j\left(\frac{N-1}{2}\right)\right] \frac{\sin\left(\frac{N}{2} f\right)}{\sin\left(\frac{f}{2}\right)}$$

$$+ 0.23 \exp\left\{-j\left[\frac{N-1}{2}\left(f \pm \frac{2\pi}{N}\right)\right]\right\} \frac{\sin\left(\frac{N}{2}\left[f \pm \frac{2\pi}{N}\right]\right)}{\sin\left(\frac{1}{2}\left[f \pm \frac{2\pi}{N}\right]\right)} \tag{9.35}$$

As noted in (9.32), the zeros of the (9.35) should be placed at the positions of the impulses that occur at $2\pi m/D$, where $m$ is an integer. The only impulse remaining must be the one at $m = 0$. The first term in (9.35),

$$\frac{\sin\left(\frac{N}{2}f\right)}{\sin\left(\frac{1}{2}f\right)} \tag{9.36}$$

defines the zeros of the rectangular part of the Hamming window, which are located at multiples of $\pi$.

From the position of the zeros

$$\frac{N}{2}f = k\pi \tag{9.37}$$

the shifting values $D$ can be derived by comparing (9.37) with the impulse positions of (9.32):

$$f = \frac{2\pi k}{N} = \frac{2\pi}{D} \tag{9.38}$$

The clear choice of $D$ is

$$D = \frac{N}{k} \quad \text{for } k = \{1, 2, \ldots, N\} \tag{9.39}$$

Furthermore, the function

$$\frac{\sin\left(\frac{N}{2}\left[f \pm \frac{2\pi}{N}\right]\right)}{\sin\left(\frac{1}{2}\left[f \pm \frac{2\pi}{N}\right]\right)} \tag{9.40}$$

defines the zeros of the cosine term in (9.34). Following a similar procedure yields

**Table 9.1**
Computed Values of $D$ for a Hamming Window of Length 128.

| $f_k$ | 64 | 32 | 16 | 8 | 4 | 2 |
|---|---|---|---|---|---|---|
| $D$ | 4 | 6 | 8 | 16 | 32 | 6 |

$$\frac{N}{2}\left(f \pm \frac{2\pi}{N}\right) = k\pi \tag{9.41}$$

where

$$f = \frac{2\pi k}{N} \pm \frac{2\pi}{N} = \frac{2\pi}{D} \tag{9.42}$$

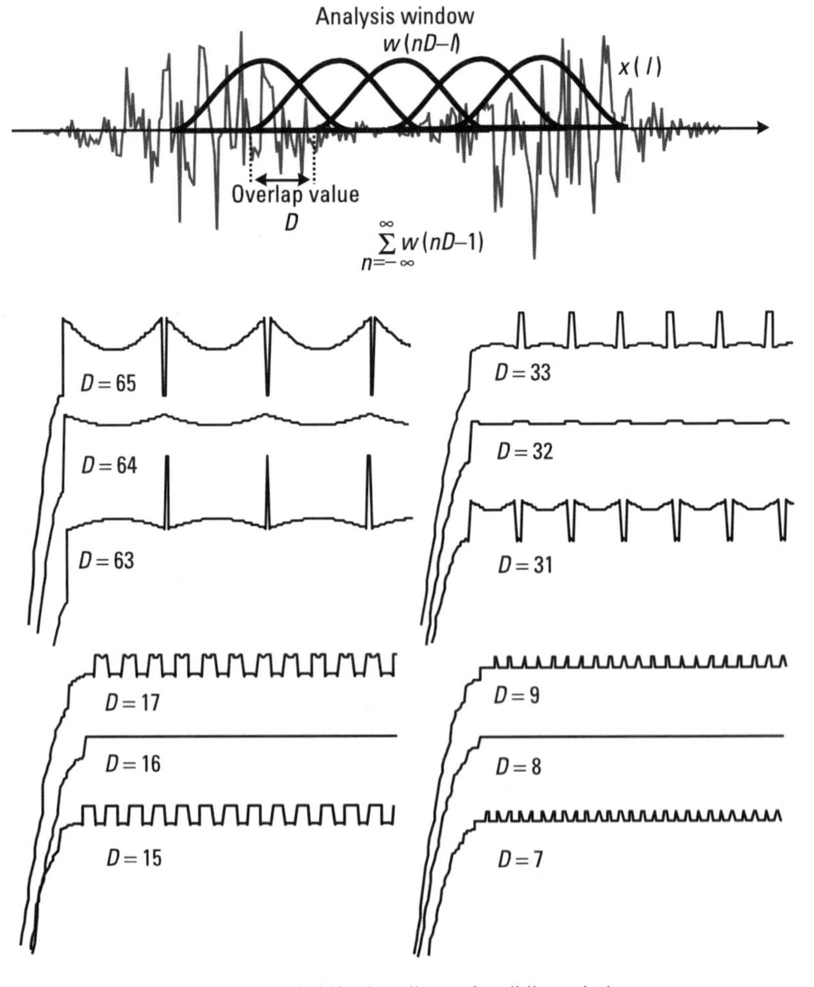


**Figure 9.14** Analysis for a series of shifted replicas of a sliding window.

resulting in a second choice for $D$:

$$D = \frac{N}{k \pm 1} \quad \text{for } k = \{2, 3, \ldots, \mathrm{N}\} \tag{9.43}$$

An example using a window length of 128 samples yields the following values:

Note that $f_k$ can only take those values because $D$ must be an integer as a consequence of the discrete implementation of the reconstruction. Furthermore, $f_k$ should start at 2, given the second condition in (9.43).

Figure 9.14 shows part of the summation in (9.27) for multiple values of $D$. The lower plots show the summation results around four estimated values of $D$ ($D = 64 \pm 1$, $D = 32 \pm 1$, and $D = 16 \pm 1$). The summation is constant when values for $D$ are used following the criteria explained in this section. Thus, near-perfect reconstruction for those values is possible.

The best possible values are obtained in accordance to (9.39) and (9.43). The finite number of samples forces the right side of (9.32) into a sinc function approximation of an impulse. The result is an error that appears as small ripples.

## 9.5 Signal Analysis and Synthesis Using Superresolution in the STFT

Resolution in the time-frequency plane becomes an important issue when nearly overlapping transient events, or long nonstationary signals are shown in the domain. A method that uses nonparametric stationary adaptive extrapolations of blocks of data, to enhance frequency-domain resolution in block-oriented implementations of linear and bilinear time-frequency representations, is presented next.

### 9.5.1 Superresolution Using Adaptive Extrapolations

A nonparametric method to extrapolate a piece of a stationary signal is used to increase resolution in the frequency direction. The adaptive weighted norm extrapolation (AWNE) algorithm proposed by Cabrera and Parks can be used to increase resolution [8]. The AWNE method works as follows for each iteration. For a given finite set of data samples $\{x(0), x(1), x(2), \ldots, x(L-1)\}$, a frequency weighted norm (the weighting is by $1/Q(f)$) is defined from an esti-

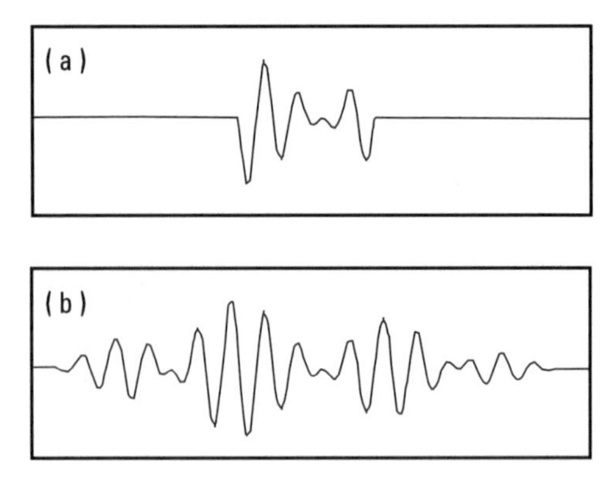


**Figure 9.15** Extrapolation example. (a) Original 31 samples; (b) extrapolation obtained using AWNE.

mated spectrum obtained from the previous extrapolation, and a minimum norm extrapolation is then computed. The process is then repeated until a fixed point of the iteration is reached. The closed form expression for the minimum norm extrapolation using the weighting $Q(f)$ is:

$$\hat{x}(n) = \sum_{i=1}^{L} b_i q(n - i) \tag{9.44}$$

where $b_i$ is an extrapolation coefficient and $q(n)$ is the inverse Fourier transform of $Q(f)$, which is an estimated spectrum. After some iterations, the resulting signal is a valid extrapolation that is a stationary extension of the given data, which is chosen (by selecting a window length in the algorithm) to be of a du-

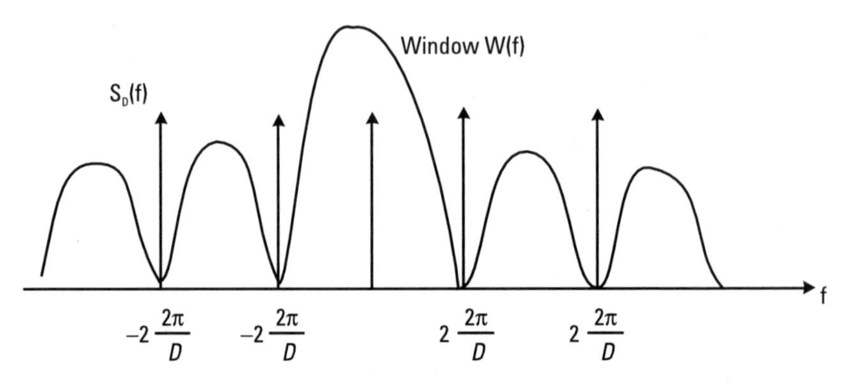


**Figure 9.16** Block diagram showing the implementation of data extensions in TFR.

ration two to five times larger than $L$, the size of the original data set [8]. An example is shown in Figure 9.15, in which a block consisting of 31 samples is extended to approximately 128 samples.

### 9.5.2 Short-Time AWNE Approach

The alternative TF representation to do TF analysis and synthesis as well as modifications for component separation is denoted the short-time AWNE (STAWNE) [9]. The enhanced version of the STFT is obtained when extrapolations are used instead of the original blocks of samples in the STFT computation [10]. Because of that, STAWNE has better frequency resolution than the STFT for the same block length $L$. Note that the increased resolution can be exchanged for better resolution in the time direction by reducing $L$, the size of the analysis block, and still have good resolution in the frequency direction.

A general scheme to enhance resolution in various TF representations is illustrated in Figure 9.16, in which adaptive extrapolations are computed for each (possibly overlapping) block to form an extrapolation history $x_H(n)$ by concatenating all of them as shown in the figure. This process can be seen to be a preprocessing step in the block-oriented computation of a TF representation (TFR), in which the samples in each original block of length $L$ are extended to length $N$ in a stationary manner (usually $N = 2L$, $3L$, or $4L$) before they are used. We denote the original and the extrapolated block of samples as $x_m(n)$ and $\hat{x}_m(n)$, respectively. The TFR is computed using the analytic signal of the extrapolation history by simply selecting a block length equal to $N$ and 0% overlapping in the selection of the next block (the block spacing $D$ is chosen equal to $N$). That strategy ensures that the full extrapolated block is used only once.

### 9.5.3 STAWNE Approach to Spectral Modification and Synthesis

The extrapolation of each block from $L$ to $N$ samples has the effect of requiring a larger DFT length $T$ (which is now required to be at least $N$) in the computation of the STAWNE, which is denoted as $STAWNE_x(n, k)$. That is, there is a need for more discrete frequency index values for each time index value in this TF representation. However, in the examples comparing results using the conventional STFT and the STAWNE, a DFT length which is the same size for both cases is selected, indicating that significant zero-padding from at least $L$ to $N$ samples will be used in the computation of the STFT.

An important characteristic of the AWNE procedure is that the original samples are present in the larger block. In practice, those samples are made to appear near its center; therefore,

$$x_m(n) = \hat{x}_m(n); \quad n = n_0, n_0 + 1, \ldots, n_0 + L - 1 \tag{9.45}$$

That fact suggests one very simple synthesis procedure consisting of simply extracting the original block of samples from each extrapolated block after the inverse DFT operation has been performed. The synthesis then proceeds the same way as for the STFT when a rectangular window $w(n)$, for which case the block spacing parameter $D$ can be any integer. Then, according to the OLA method, the reconstruction can be achieved by adding shifted versions of $\hat{x}_m(n)$:

$$Cx(l) = \sum_m x_m(l - (m - 1)D) \tag{9.46}$$

where $C$ is easily shown to be equal to the Fourier transform of the rectangular window at $w = 0$, which is $W(0) = L$ divided by $D$ [11, 12].

Another possible synthesis procedure from STAWNE, which has been found to work reasonably well, assumes that the extrapolated block is actually a windowed version of the original signal $x(n)$. That can be expressed as

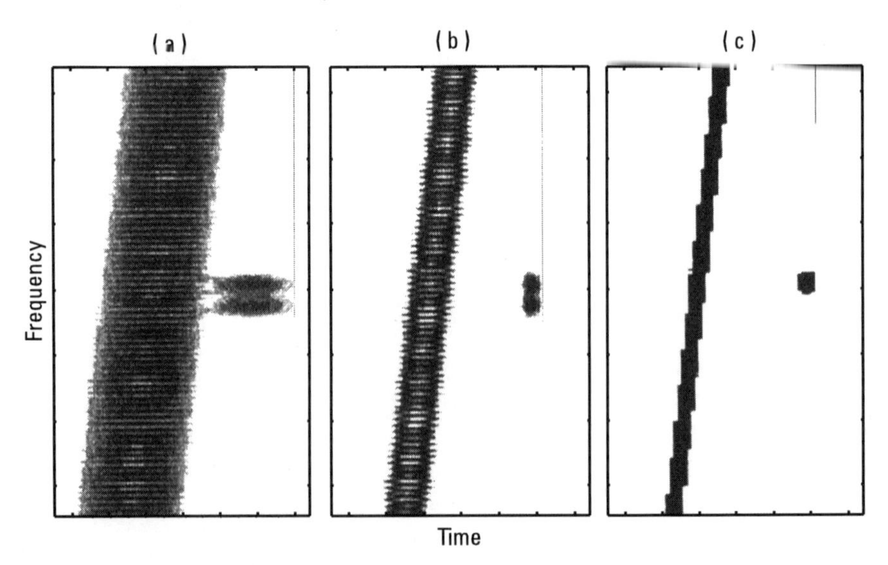


**Figure 9.17** (a) Magnitude of the STFT of a multicomponent signal corresponding to two frequency-modulated sinusoids and two modulated decaying exponentials. Good time localization results at the expense of resolution in frequency. (b) Magnitude of the STAWNE representation. Note how temporal information is kept while resolution along the frequency axis is improved. (c) The two pass regions used to separate two desired components.

$$\hat{x}_m(n) \approx x[n + (m - 1)D]h(n) \tag{9.47}$$

where $h(n)$ is an unknown but well-behaved window of length $N$ that resembles the Tukey window (see Figure 9.15). The synthesis can then be performed to within a constant, assuming that $h(n)$ is a rectangular window, as follows:

$$C'x(l) = \sum_m \hat{x}_m[l - (m - 1)D] \tag{9.48}$$

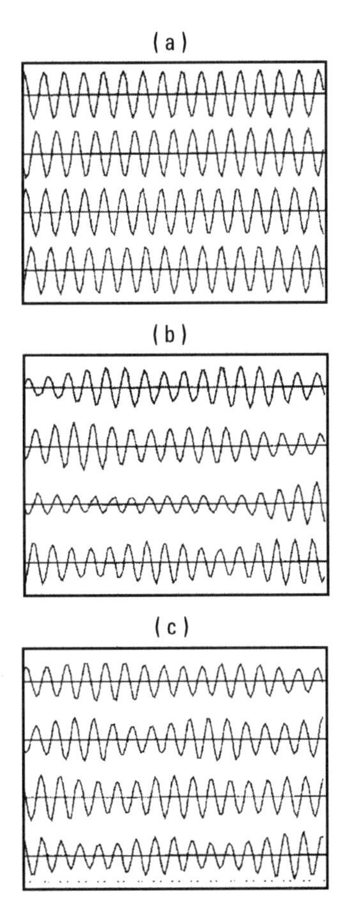


**Figure 9.18** (a) 500 samples of the original linear frequency-modulated component; (b) synthesis of the FM component from modified STFT obtained using the appropriate part of the pass-region in Figure 9.17; (c) synthesis obtained from the modified STAWNE after using the same mask in (b).

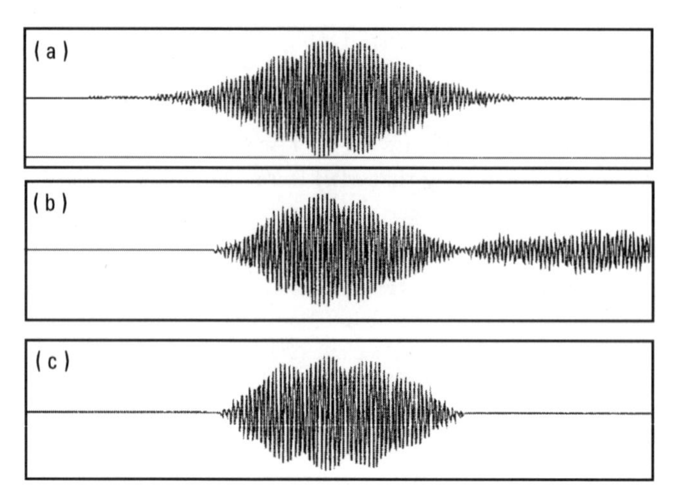


**Figure 9.19** (a) Transient signal corresponding to one of the modulated symmetric decaying exponentials; (b) synthesis from modified STFT; (c) synthesis of the transient component from modified STAWNE.

where the constant $C'$ is not known exactly. The filterbank interpretation [1, 13] is to substitute $x(n)$ for $x_H(n)$ at the input of the synthesis filters with some minor modifications.

### 9.5.4 Example of Component Separation Using STFT and STAWNE

An example of editing or filtering components in the TF domain is shown in Figure 9.17. The test signal consists of two linear FM chirps and two transients (modulated symmetric decaying exponentials). The signal separation is done by using both TF representations, the STFT, in Figure 9.15(a), and the STAWNE, in Figure 9.15(b), with the same discrete time and frequency index spans for both. A look at the magnitude plots for both representations clearly shows that the STAWNE has better frequency resolution (the block size used is the same) while still preserving the same time localization. The STAWNE with its higher resolution is used to define the TF filter that selects the two desired components. (Chapter 10 presents more details on TF filters.)

To appreciate the results, we need to consider how well the components are extracted from the total signal. That is shown in Figures 9.18 and 9.19 for a chirp and a transient component, respectively. In both cases, the synthesis of the signal is performed after the STFT or the STAWNE representation has been modified by being multiplied by the mask of 1 or 0 values in the discrete time-frequency plane. We can clearly see the superiority of the STAWNE editing process by comparing the ideal component in Figures 9.18(a) and 9.19(a) with the results of the separation in Figures 9.18(b) and 9.19(b) using STFT, and Figures 9.18(c) and 9.19(c) using STAWNE.

## References


[1] F. Halawatsh, and G. F. Boudreaux-Bartels, "Linear and Quadratic Time-Frequency Signal Representations," *IEEE Signal Processing Magazine*, pp. 21–67, April 1992.

[2] L. Cohen, and C. Lee, "Instantaneous Bandwidth," in *Time Frequency Signal Analysis*, B. Boashash (Ed.), Wiley Halsted Press, Melbourne, Australia, 1992, pp. 98–117.

[3] Marple, "Are Quadratic Time-frequency Representations Really Necessary?," *Proc. ICASSP96*, Vol. 5, pp. 2575–2579, Atlanta Georgia, May 1996.

[4] Cohen, L., "Introduction: A Primer on Time-Frequency Analysis," in *Time Frequency Signal Analysis*, pp. 3–42, B. Boashash (Ed.), Melbourne, Australia. Wiley Halsted Press. 1992.

[5] Y. Zhao, L. Atlas, and R. Marks, "The Use of the Cone-Shaped Kernels for Generalized Time-Frequency Representations of Nonstationary Signals," *IEEE Trans. on Signal Processing*, Vol. 7, 1990, pp. 1084–1091.

[6] H. Nawab, and T. F. Quatieri, "Short-Time Fourier Transform," In Chapter 6 of *Advanced Topics in Signal Processing*, J. S. Lim, and A. V. Oppenheim (Eds.), Prentice Hall, New Jersey, 1988, pp. 289–337.

[7] F. J. Harris, "On the Use of Windows for Harmonic Analysis with the Discrete Fourier Transform," *Proc. of the IEEE*, Vol. 66, No. 1, January 1978.

[8] S. D. Cabrera and T. W. Parks, "Extrapolation and Spectral Estimation With Iterative Weighted Norm Modification," *IEEE Trans. on Signal Processing*, Vol. 39, No. 4, April 1991, pp. 842–851.

[9] S. D. Cabrera, B. C. Flores, G. Thomas, and J. Vega-Pineda, "Evolutionary Spectral Estimation Based on Adaptive Use of Weighted Norms," *Proc. SPIE Advanced Signal Processing Algorithms, Architectures, and Implementations IV*, Vol. 2027, July 11–13, San Diego CA, 1993, pp. 168–179.

[10] G. Thomas, and S. D. Cabrera, "Resolution Enhancement in Time-Frequency Distributions Based on Adaptive Time Extrapolations," *Proc. IEEE-SP Int. Symp. on Time-Frequency and Time-Scale Analysis*, Philadelphia, PA, October 1994, pp. 104–107.

[11] A. Papoulis, The Fourier Integral and its Applications, McGraw-Hill, 1962.

[12] J. B. Allen, "Short Term Spectral Analysis, Synthesis, and Modification by Discrete Fourier Transform," *IEEE Trans. on Acoust. Speech Sig. Proc.*, June 1977, pp. 235–238.

[13] M. R. Portnoff, "Time-Frequency Representation of Digital Signals and Systems Based on Short-Time Fourier Analysis," *IEEE Trans. on Acoust. Speech Sig. Proc.*, Vol. 28, No. 1, February 1980, pp. 55–68.

# 10

# Selective Motion Compensation

The preceding chapters discussed the analysis and synthesis techniques to be used with target signatures. This chapter proposes a selective motion compensation methodology based on the techniques.

## 10.1 Definition of the Time-Varying Filter

The time-varying filter takes the form of a binary mask, which is zero-valued over regions of no interest in the TF plane and has a unit value in areas where target signatures are concentrated. The mask is defined as [1]

$$M(t_n, f_k) = \begin{cases} 1, & \text{for } t_n, f_k \in \Re \\ 0, & \text{for } t_n, f_k \notin \Re \end{cases} \tag{10.1}$$

where $\Re$ represents the target signature region of interest.

Time-varying filtering can be done by multiplying the STFT of the multiple target signatures with this mask. The result yields a modified STFT (MSTFT ) described as

$$MSTFT(t_n, f_k) = STFT_x(t_n, f_k)M(t_n, f_k) \tag{10.2}$$

where $x$ represents a superposition of $L$ different target signatures $x_i$. That is,

$$x \sum_{i=1}^{L} x_i$$

so that

$$STFT_x(t_n, f_k) = \sum_{i=1}^{L} STFT_x(t_n, f_k)M_i(t_n, f_k) \tag{10.3}$$

where $STFT_{x_i}(t_n, f_k) = STFT_x(t_n, f_k)M_i(t_n, f_k)$ since the STFT is a linear transformation and

$$\sum_{i=1}^{L} M_i(t_n, f_k) = 1 \quad \forall t_n, f_k \in \Re_1, \Re_2, \ldots, \Re_L$$

The binary filter $M_i(t_n, f_k)$ covers the $i$th signature support region. The synthesized signature obtained using the inverse of $STFT_{x_i}$ allows the motion compensation of each signature separately.

## 10.2 Evaluation of Time-Frequency Filtering

This section examines the effect of filtering in the time-frequency domain, focusing on the effects introduced to the phase of the scatterer response when it is isolated using the filtering. The multicomponent signal used is modeled as

$$x(n) = \sum_i A_i \exp[j\psi_i(n)] = \sum_i A_i \exp\left[j\pi\left(2f_i\frac{n}{f_s} + \nu_i\frac{n^2}{f_s^2} + \phi_i\right)\right] \tag{10.4}$$

and the instantaneous frequency associated with the $i$th scatterer in the image is estimated and compared to the estimated frequency obtained for the case where only that scatterer is present by itself. Deviations caused by those two computations are presumed to affect the final compensation of an image regardless of which compensation technique is used [2].

From the instantaneous frequency (IF) expression in (9.2), the IF for the continuous model of (10.4) for a single scatterer is defined as

$$f_I(t) = f_i + \nu_i t \tag{10.5}$$

where $f_I(t)$ is directly related to the motion parameters. Any phase changes that occur during the TF filtering and synthesis steps affect the estimation of those parameters and consequently the compensation process. Different IF estimators are discussed next; a more complete evaluation of these estimators is presented by Boashash in [3].

### 10.2.1 IF From the Smoothed Central Finite Difference Method

The central finite difference (CFD) estimator is defined as an approximation of the slope of $f_I(t)$ around time index $n$:

$$\hat{f}_I(t) = \frac{1}{4\pi}(\psi(n+1) - \psi(n-1)) \tag{10.6}$$

This estimator is very susceptible to noise. To reduce the large variance presented at low SNRs, a smoothed version of the estimator (which will be called SCFD) has been proposed by Kay [4], in which a smoothing weighting function of the form:

$$w_n = \frac{\frac{3}{2}N}{N^2 - 1}\left\{1 - \left[\frac{n - \left(\frac{N}{2} - 1\right)}{\frac{N}{2}}\right]^2\right\} \tag{10.7}$$

is used on a set or $N$ samples in conjunction with (10.6), yielding the single sinusoidal frequency estimator:

$$\hat{f}_0 = \frac{1}{2\pi}\sum_{n=0}^{N-2} w_n[\psi(n+1) - \psi(n-1)] \tag{10.8}$$

To track a linear chirp, the window is shifted one sample at a time over the entire set of samples, as described in [3], and a new frequency estimation is stored for each shift. The final record is then the resulting estimated IF $\hat{f}_I(n)$.

### 10.2.2 IF From Linear Prediction and Signal Modeling

This method models a time-varying process as an autoregressive (AR) model of order $P$ at each time index $n$. In order to estimate the $P$ coefficients, the linear predictor arrangement depicted in Figure 10.1 is used. To update the filter taps, the least mean square (LMS) algorithm is used. The LMS algorithm tries to minimize the error between the desired signal and the output: $e(n) = d(n) - y(n)$. The minimization is achieved via the partial derivative of the cost function with respect to the filter coefficients [5]

$$\nabla J_n = \frac{\partial J}{\partial \underline{a}_n} = \frac{\partial E[e^2(n)]}{\partial \underline{a}_n} \tag{10.9}$$

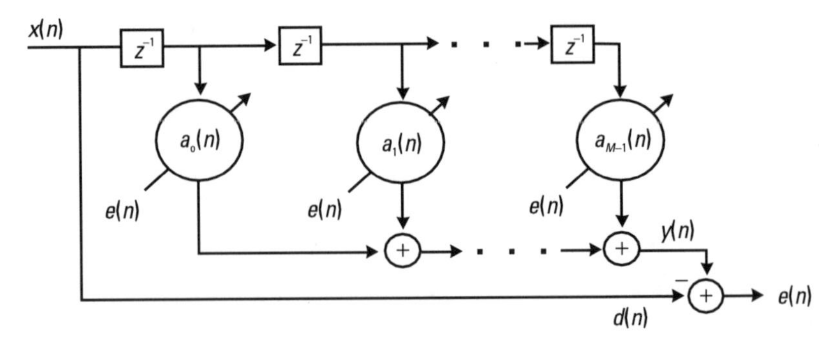


**Figure 10.1** Linear predictor model for adaptive estimation of the instantaneous frequency.

where $E$ is the expected value operator.

The updated coefficients are found after estimating that Laplacian as

$$\hat{\nabla} J_n = \frac{\partial e^2(n)}{\partial \underline{a}_n} \tag{10.10}$$

As indicated in [5], the coefficients at the next time instant, $n + 1$ are found according to

$$\underline{a}_{n+1} = \underline{a}_n + \alpha e(n) \underline{x}_n \tag{10.11}$$

where $\alpha$ is an adaptation constant that defines the rate of conversion of the algorithm.

As a new set of coefficients, $\underline{a}_n$ is found, the AR power spectrum is calculated according to [6]

$$S_{AR}(f, n) = \frac{\sigma_v^2}{\left| 1 + \sum_{k=1}^{P} a_k(n) \exp(j2\pi fk) \right|^2} \tag{10.12}$$

where $\sigma_v^2$ indicates the noise power. The peak position determines the IF.

Because a monocomponent signal is isolated via TF filtering, a faster approach to IF estimation using that method is possible for this case [3]. The AR model must have only one narrow band component (one pole), and the IF can be approximated as:

$$\hat{f}_I(n) = \arg\{a_1(n)^*\} \tag{10.13}$$

The single-pole model saves the computational time of calculating the spectrum and finding the peaks.

### 10.2.3 Computer Simulations

As noted by Boashash [7], TF filtering can be applied to reduce the variance of an IF estimator. Here, the filtering is done using the STFT, and the consequences of doing that are studied not only for low-SNR cases but also for those with higher SNR values. Simulations start at −5 dB SNR because from that value on, the multicomponent signal is distinguishable from noise using the STFT. Once the filtering has taken place, the IF estimators discussed in the previous sections are applied to the reconstructed signal.

Each method is compared against the Cramer-Rao bound (CRB) formulated by Wong and Jin [8], which is expressed for continuous time as

$$E[e_l] \geq \frac{\sigma_v^2}{2AT_l}\left\{\frac{\int_{-T_l/2}^{T_l/2} f_l^2(t)\,dt}{\int_{-T_l/2}^{T_l/2} t^2 f_l^2(t)\,dt}\right\} \tag{10.14}$$

where $T_l$ is the duration of the signal.

Results computing 200 realizations of monocomponent signals and filtered multicomponent signals buried in noise are shown in Figure 10.2. The MSE between the true IF and the estimated IFs are plotted against the SNR values. It can be seen that the MSE is reduced by filtering when SNR levels are low. Improvements around 12 dB such as the ones found by Boashash and White [9] are also obtained with use of STFT filtering. For the SCFD and the LMS algorithms, the results between single components and filtered components when the SNR is higher are very similar. The results greatly motivate the use of TF filtering before those IF estimators are applied. A drawback on the LMS method is that it requires us to choose a parameter (the adaptation constant $\alpha$) that can affect the results considerably.

Another method to estimate the IF is to look at the peaks of TF representations. For each instant $t$ in the two-dimensional representation $T_x(t, f)$, the location of the frequency peak is stored, making that the estimate of the IF. For the WVD and the CWD estimators, the IF estimation is enhanced for low SNRs, but it seems that the filtering introduces some bias in the estimation, making the results worse for better SNR values.

The decomposition of a multicomponent signal using TF filtering not only reduces the variance of the IF estimators at low SNRs, it also allows the

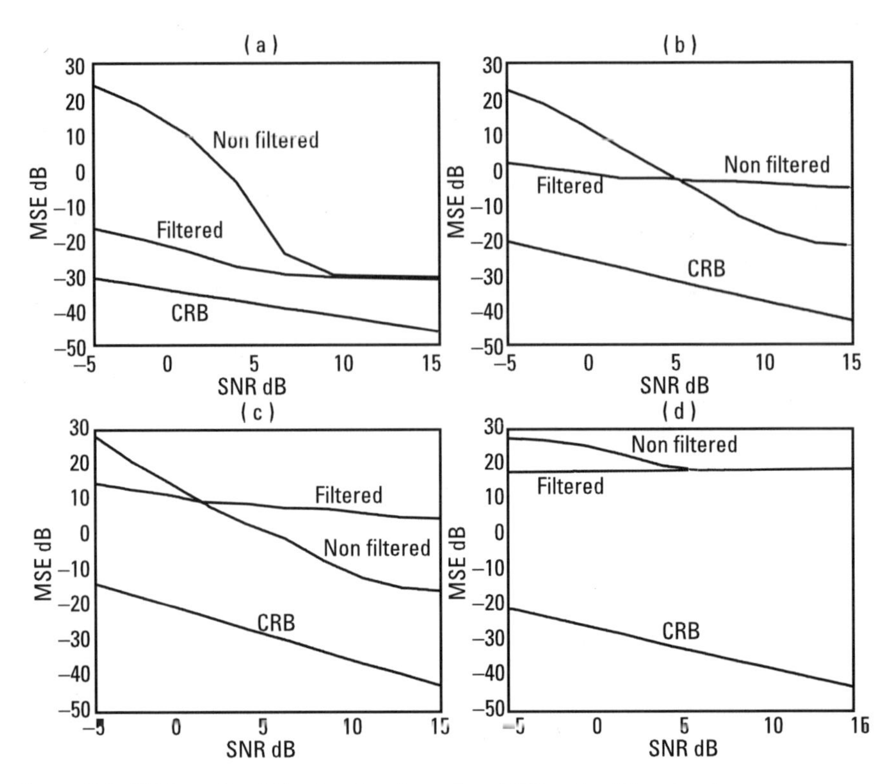


**Figure 10.2** MSE results of IF estimation for various SNRs using filtered and monocomponent signals. (a) Using the SCFD; (b) using the peaks of the WVD; (c) using the peaks of the CWD, d) using the LMS method. It can be seen in all four cases that the MSE at low SNRs is lower when using the TF filtering.

use of methods that are based on monocomponent signals. The methods that can be used due to the presence of only one component are the SCFD and the faster approach based on the adaptive method using the LMS algorithm. Besides the benefit of using TF filtering for selective motion compensation purposes, filtering can also be used as a denoising technique that can improve the compensation results on a SAR/ISAR image.

## 10.3 Binary Mask Formation

Two steps are required to define the binary mask. First, a binary filter is obtained from the squared absolute value of the STFT, also known as the spectrogram (SG) of the signal. Second, the filter is enhanced by morphological operations. Finally, each filter component to used for selective compensation is separated via binary filter labeling.

The binary mask is defined using the two following hypotheses [10]:

- $H_0$: Only noise is present.
- $H_1$: A target signature return and noise are present.

Given a CFAR, a decision is made whether a particular value on the TF plane comes from $H_0$ or $H_1$. If noise characteristics change from frame to frame, CFARs can be achieved by updating the parameters of the PDF. Once the threshold has been computed for a CFAR, the threshold is applied to obtain a preliminary pattern for the time-varying filter.

The SG regions dominated by noise can be exploited to approximate the noise distribution to a particular PDF. For this work, we chose the Weibull PDF and followed the same procedure described in Section 6.2.

A threshold is computed according to 6.31. Using that threshold, $H_1$ is rejected if a pixel from the SG is below $\eta$ and a zero value is assigned to the filter at that point. Likewise, $H_0$ is rejected if the pixel value is above $\eta$ and a 1 is assigned to the filter at this position. Evaluating the entire SG forms the preliminary binary filter.

## 10.4 Filter Enhancement

To reduce noise and isolate signatures effectively, a low CFAR is selected during the formation of the binary filter. Thus, the TF filter at this point has spurious peaks and small gaps within a desired TF support region. Consequently, two morphological operations are desirable to enhance the TF mask [11].

First, a majority operation is applied for spurious pixels present on the TF filter. This operation consists of assigning a value of zero to a pixel if, within 3-by-3 neighborhoods of the binary filter's pixel, five or more pixels are zero. Thus, any of the situations depicted in Figure 10.3 yields a zero-valued 3-by-3 area image.

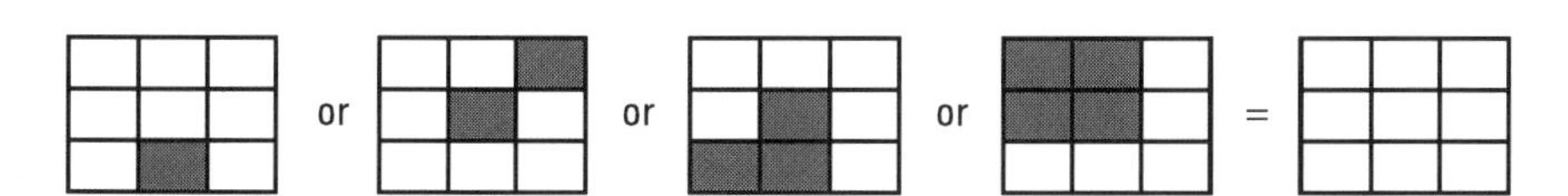


**Figure 10.3** Example of majority operation.

If a CFAR is to be defined to skip this operation, a threshold using (6.31) must be defined so that, at the most, only one false alarm is shown as a spurious pixel on the binary filter. That lets us define a lower boundary for the CFAR, because a very low CFAR would yield a higher threshold that would make the pass region of the time-varying filter miss target signature regions.

Let us define the CFAR as [12]

$$\text{CFAR} = \frac{1}{N_{tot}} \frac{t_{int}}{t_{fa}} \tag{10.15}$$

where $N_{to}t$ indicates the total number of time-frequency bins in the spectrogram, $t_{int}$ is the integration time, and $t_{fa}$ is the average time between false alarms. If, on the average, only one black pixel is to be shown in a noisy image, the time ratios $t_{int}/t_{fa}$ must be equal to 1, thus defining a lower bound for the CFAR value. That bound will define a threshold that is high enough, but more than one false alarm is desirable so that the threshold will be lowered to cover the entire target signature region.

To check the control of the average number of false alarms, 200 experiments consisting of 2,000 samples were performed. Figure 10.4 shows a number of experiments where Weibull-distributed noise samples are generated to simulate different numbers of pixels that appear on either a range-profile history or spectrogram images consisting of noise only. A threshold is computed using (6.31), and the number of false alarms desired is specified according to (10.15), that is, the number of black pixels in the image. On average, the desired number of false alarms (one and four for the two examples) for the calculated CFAR is obtained.

A closing operation between the original filter and an eight-connected structuring element $B$ is used to fill the gaps of filter $A$ (thus forming a region of connected pixels with value of 1), as illustrated in Figure 10.6. This operation is implemented by dilating and eroding a binary image, that is, by adding and removing eight connected pixels around the binary component. As indicated in [11], the application of these two operations results in the closing operation. The results of using the majority and closing operations are shown in Figure 10.5(b).

Monte Carlo simulations show that the closing operation yields no better results in terms of ISAR image entropy values than the use of the majority operation alone, as shown in Figure 10.7. Here, the filter is defined by using the threshold operation alone. Another filter definition consists of applying the majority operation. Finally, a filter is defined by applying the majority operation followed by closing.

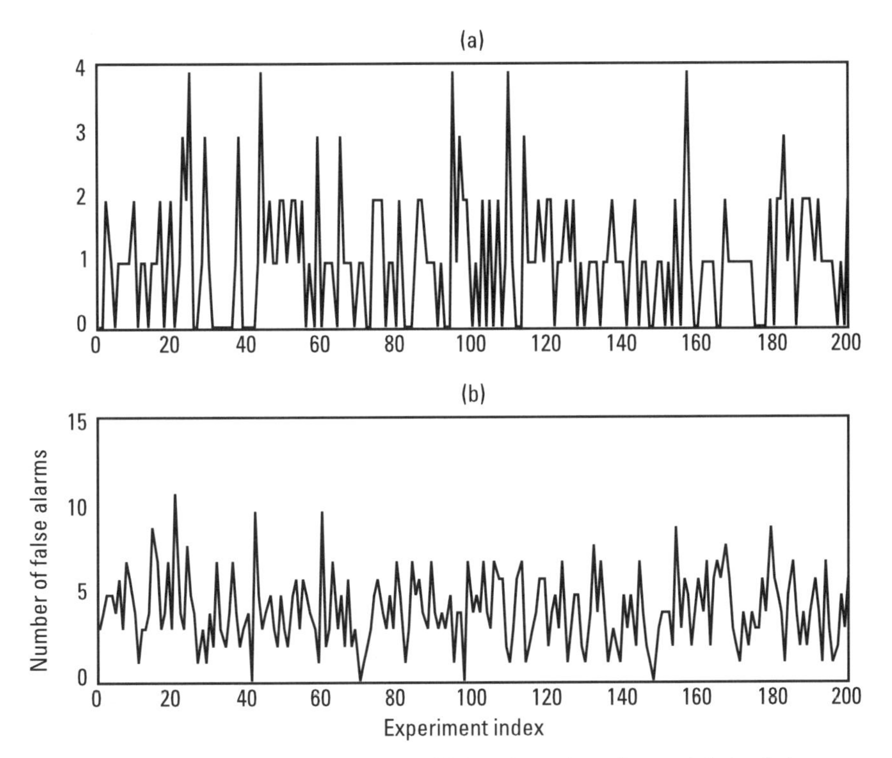


**Figure 10.4** Simulation of average number of false alarms in an image: (a) simulation corresponding to one false alarm; (b) case for four false alarms.

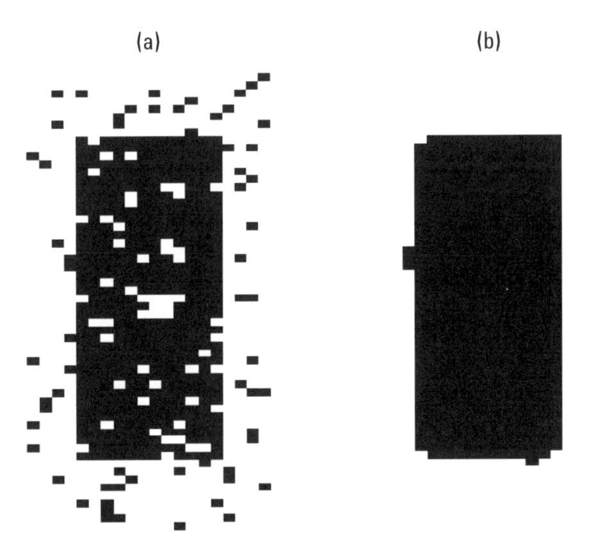


**Figure 10.5** Morphological operations: (a) original rectangular filter: (b) filter after majority and closing operation of original rectangular filter.

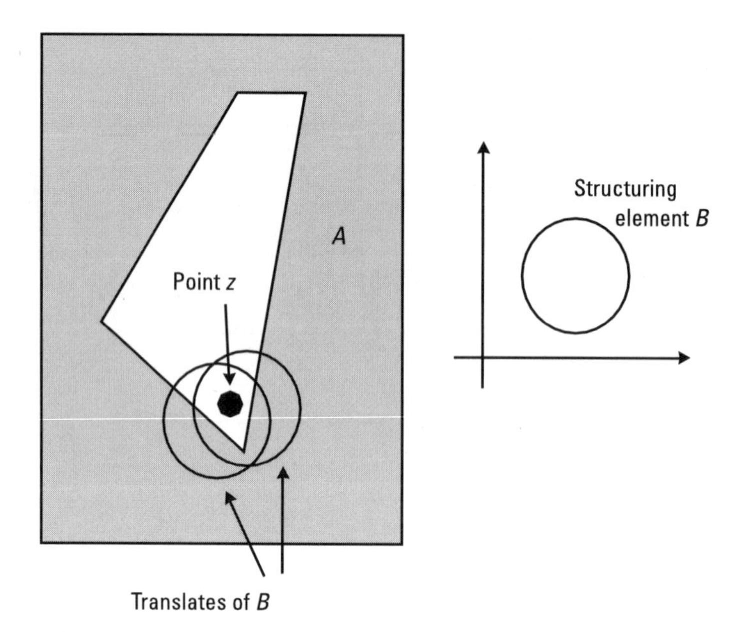


**Figure 10.6** Closing operation in blob $A$. Here, $z$ is an element of the filter if and only if $(B + y) \cap A \neq 0$, for any translate $B + y$ containing $z$. The effect is to close the hole in filter $A$.

Furthermore, reconstruction is deteriorated by filling the gaps, because noise prevails at those time-frequency bins. Thus, we considered the closing morphology as an optional operation. It is included to form a binary filter that covers the target signature TF trajectory regardless of the amplitude scintillation and noise effects.

## 10.5 Filter Labeling

The next objective is to decompose a binary filter into several filter components, each a separate binary blob (corresponding to a single target signature). The labeling method used for this purpose consists of scanning the entire binary filter from left to right and top to bottom using a five-pixel mask. The mask is illustrated as the shadow portion in Figure 10.8. Whenever the darker pixel of the mask finds a 1 (black pixel), a decision is made and a label is assigned to that pixel according to the situations listed in Table 10.1 [13].

Afterward, the entire filter is labeled again in ascending order so that the labels start with the integer one and the last label corresponds to the total number of target signatures.

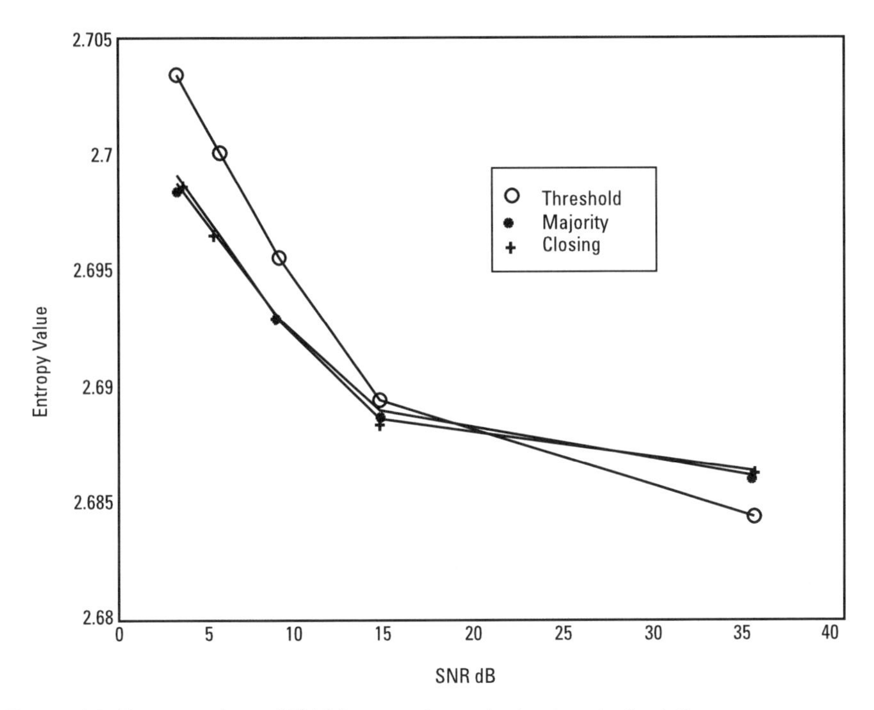


**Figure 10.7** Entropy values of ISAR image using a single-aircraft simulation.

The upper part of Figure 10.8 shows an example when situation 3 is encountered, that is, two different labels fall within the analysis mask. The mask consists of the five gray pixels. When the bottom right pixel (shown darker in the image) encounters a black pixel, a decision must be made. The lower part of Figure 10.8 shows the decision made. From the two labels within the mask, the decision is to change all the connected pixel labels to the maximum label of the possible two.

**Table 10.1**
Labeling Method of Binary Filters

| Situation | Action to Take |
|---|---|
| 1. The surrounding pixels are all zero. | The label is assigned as a new one. |
| 2. There is another pixel with a label. | Assign that label to the new pixel. |
| 3. There are two pixels with different labels. | Assign the label with highest number and change the rest of the pixels labels in the component found. |

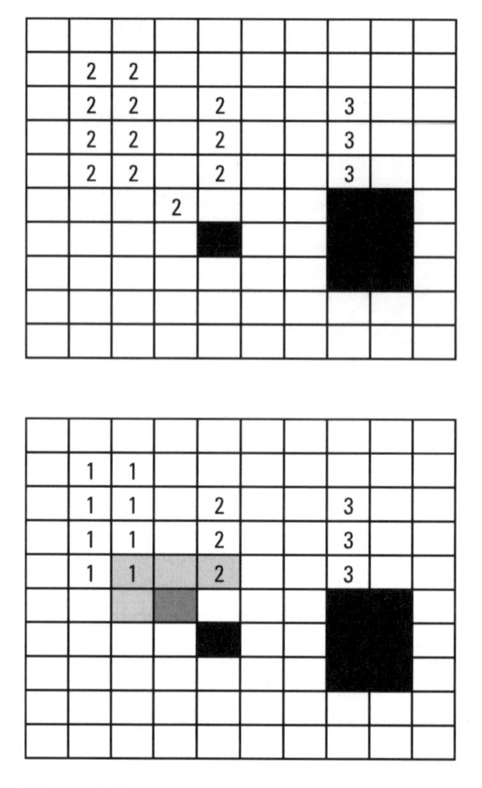

**Figure 10.8** Example of filter labeling.

## 10.6 Algorithm Description and Examples

The steps for imaging multiple targets are outlined next and summarized in Figure 10.9. Two real aircraft signatures were added in the spatial frequency domain and processed to illustrate the steps.

1. Having collected samples from the depicted scenario with multiple moving targets, a SG is obtained using the STFT. Although the samples are collected in frequency, the STFT can be used, because the minus sign in the exponential term in (9.19) can be interchanged between the forward and inverse operations and filtering can be performed. Figure 10.10 shows the SG for the superimposed signatures. Notice the similarity between the SG and the target's range profile history. Also not how resolution in the STFT is not as good as in the range profiles image, which is because the window length is half the burst length. The overlapping (half the window length) introduces more time bins in the STFT image.

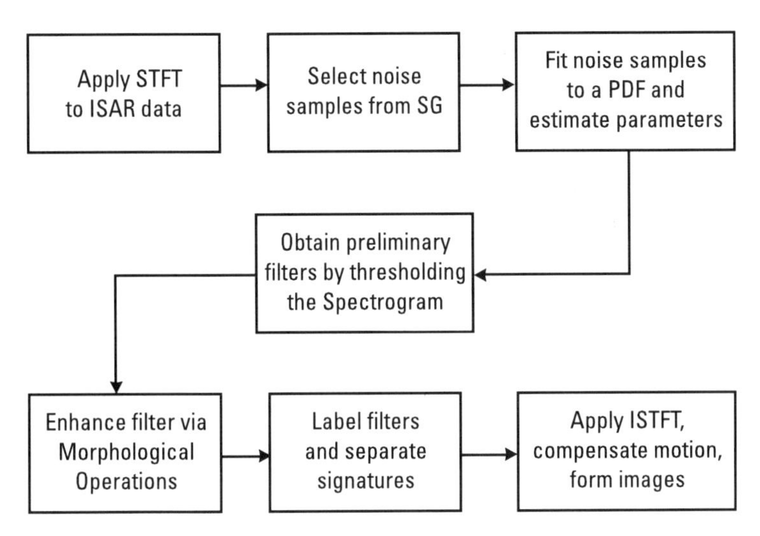


**Figure 10.9** Selective motion compensation steps.

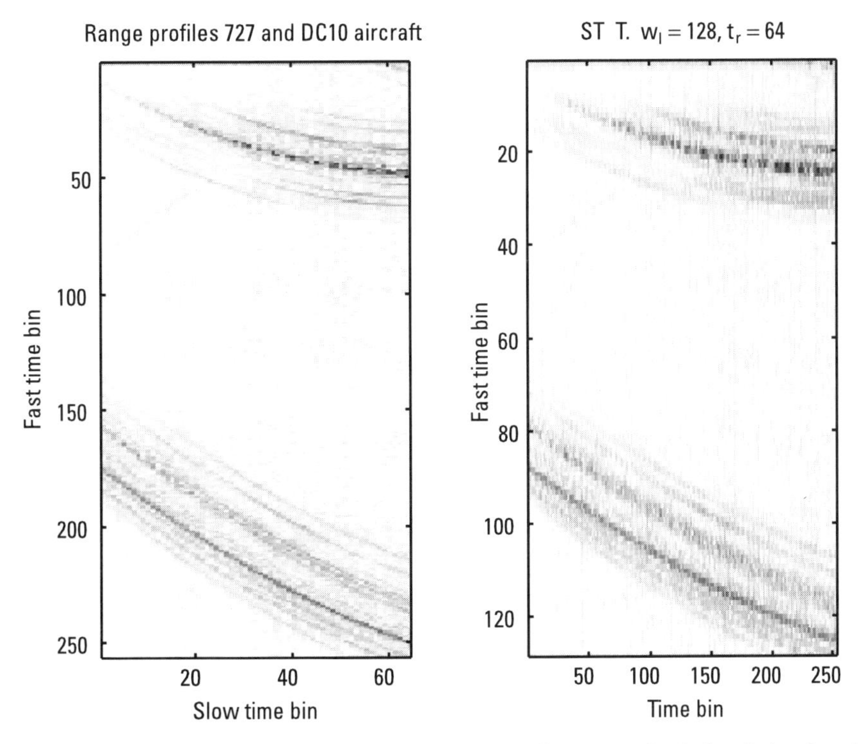


**Figure 10.10** Range profiles and STFT of superimposed signatures. Analysis window length ($w_l$) is 128 samples. Window overlap value ($t_r$) is 64.

2. To obtain a CFAR threshold, a set of noise samples is selected from the original TF. The Weibull parameters $a$ and $b$ are then estimated using Menon's method. The threshold $\eta$ is used to specify which bins correspond to an aircraft signature and which correspond to noise only. Figure 10.11 shows the histogram and the fitted PDF when TF filtering is used. The difference in amplitude between both histograms comes from the use of a different number of samples.
3. At this point, the binary filters need to be enhanced utilizing the majority and closing operations. After all spurious peaks and gaps have been removed, each aircraft signature is identified, separated, and labeled. Figure 10.12 shows the binary filter after thresholding the SG and the enhanced and separated filters obtained via the morphological operations. The order in which the operations must be taken as shown in Figure 10.12, is clockwise.

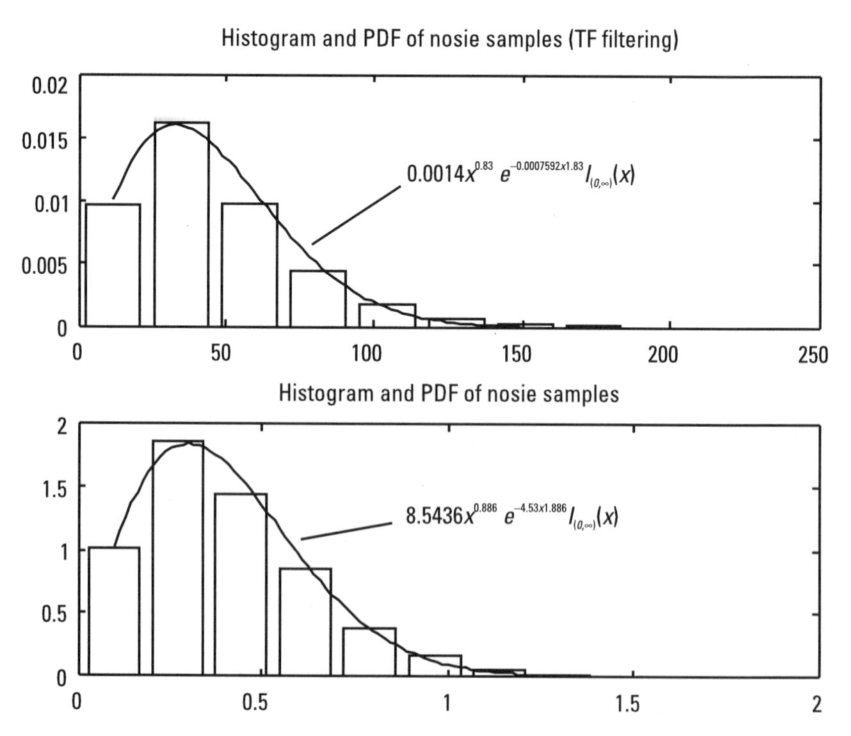


**Figure 10.11** Histogram and PDF of noise samples using TF and burst filtering.

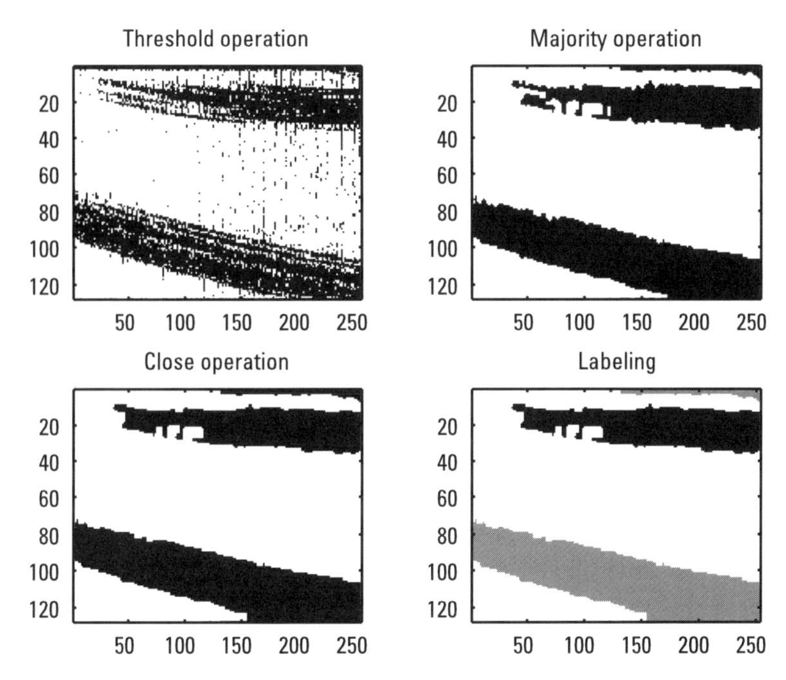


**Figure 10.12** Different stages of filter enhancement.

4. The filtering takes place when the TF representation of the superimposed signature in step 1 is multiplied by the filters obtained in step 3. Signal reconstruction is then performed so the initial velocity and acceleration of each target can be estimated using the minimum entropy method.
5. A complex exponential is formed using the best possible velocity and acceleration estimates for each target. The compensation of each target signature is done separately. An image is obtained for each target by applying a two-dimensional Fourier transform to its compensated signature. The final images for each aircraft are shown in Figure 10.13.

When the TF representation of two signatures happen to cross each other, the filtering operation yields blurred images, as shown in Figure 10.14. That occurs even when intrusive prominent scatterers are removed, as shown in Figure 10.15. Therefore, a necessary condition for obtaining focused images is that the signatures do not cross in the TF domain.

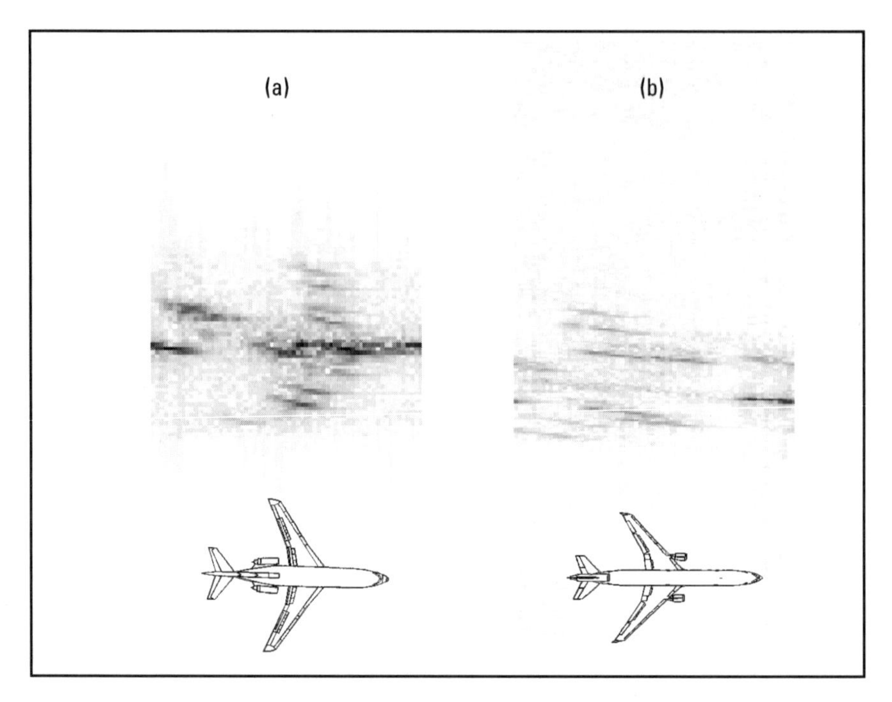


**Figure 10.13** Selective compensated images of (a) 727 aircraft and (b) DC10 aircraft.

Another imaging possibility is to obtain a complete focused image by adding the compensated signatures into a single sample record. That is possible because the STFT is a linear operation. The following simulated example illustrates the possibility in which two simulated aircraft are imaged in a single frame.

Figure 10.16 shows an ISAR image of two aircraft when no motion compensation is performed. As expected, neither aircraft is discernible from the image.

Figure 10.17 shows the range profile history of both aircraft. The different slopes are related to the different kinematics motion parameters.

The two time-frequency filters used for separation are shown in Figure 10.18. No noise was added because the purpose is to demonstrate the single-frame example. Thus, the definition of the time-varying filter technique was not applied. A single threshold defined the masks in this case.

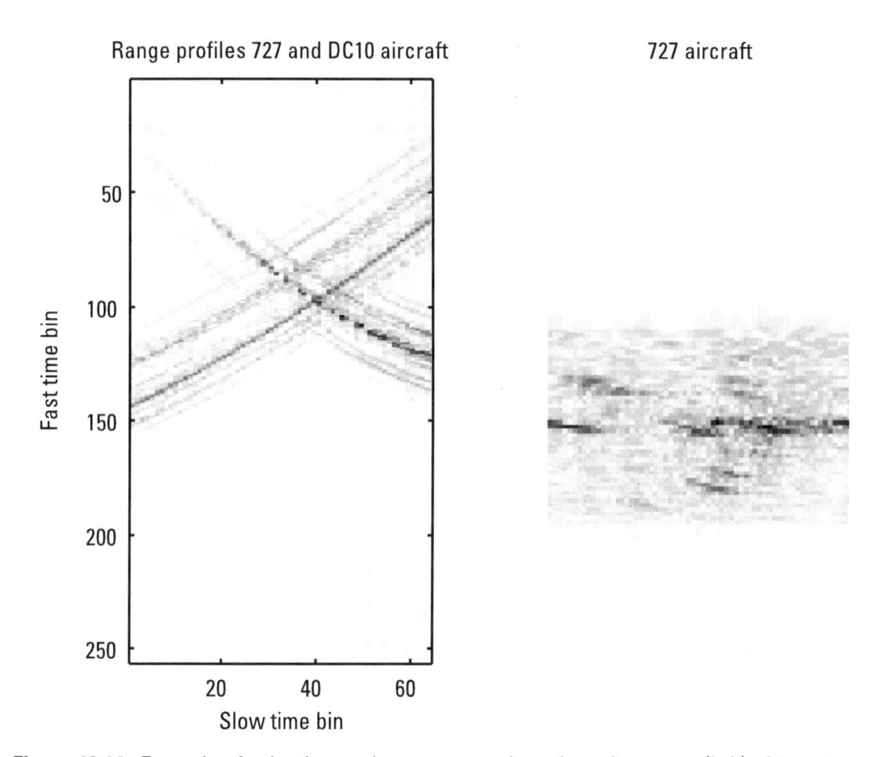


**Figure 10.14** Example of selective motion compensation when signatures (left) of targets are superimposed. Blurring appears on the final image (right).

If the separation is not performed and the kinematics parameters of one of the aircraft is used, the compensation will provide only one focused object. This can be seen in Figures 10.19 and 10.20. Note how in the range profile history, the compensation eliminates one of the signatures tilting, but affects the other one as well without compensating the object. This is clearly seen in the final image obtained, depicted in Figure 10.20, where only one target appears focused.

Finally, if the time-varying filters are used and the two signatures are compensated separately, the compensated signatures can be added and generate a complete focused image as shown in Figure 10.21.

We would like to note that another TF filtering technique using another TF representation and synthesis could be used. Although the ISAR data form a two-dimensional image, the collected echoes form a time series that can be analyzed, transformed, and synthesized using a two-dimensional time-frequency

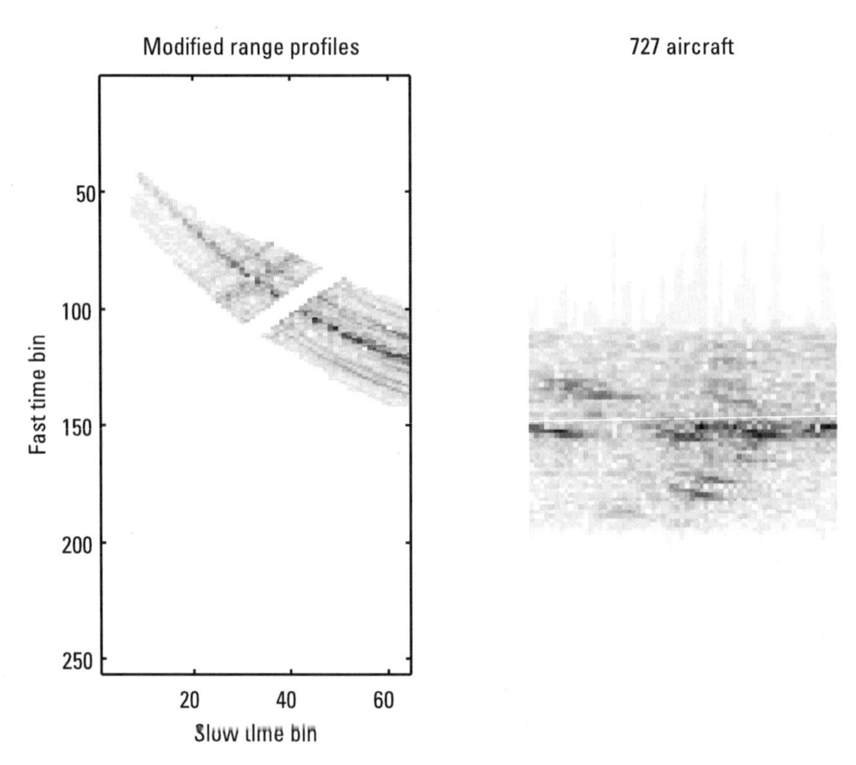


**Figure 10.15** Modification of the 727 aircraft filter to suppress superimposed prominent scatterer of DC10 signature. Blurring on the 727 final image is evident.

technique. Figure 10.22 shows two TF plots of the samples collected in Figure 1.1. Figure 10.22(a) shows the TF plot obtained using the STAWNE (technique introduced in chapter 9), and Figure 10.22(b) was obtained using the STFT. Only the samples from the range profiles corresponding to the rotating antenna of the vessel were used. As it can be seen from the contour plots in Figure 10.22, the better resolution obtained with the STAWNE approach facilitates the definition of TF filters that separate the returns from the rotating antenna. The TF filtering using the STAWNE eliminated the blurring part of the image, as it can be seen in Figure 1.1.

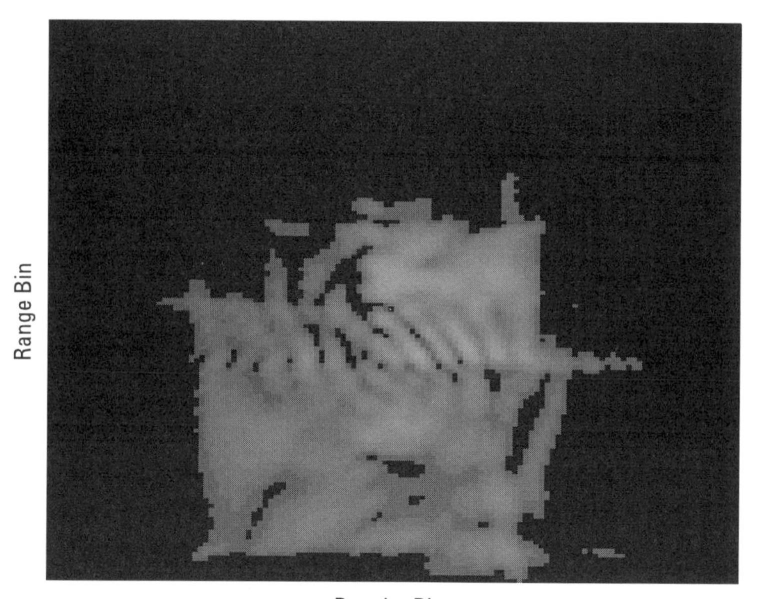


**Figure 10.16** ISAR image without motion compensation.

**Figure 10.17** Range profile history of two uncompensated aircraft.

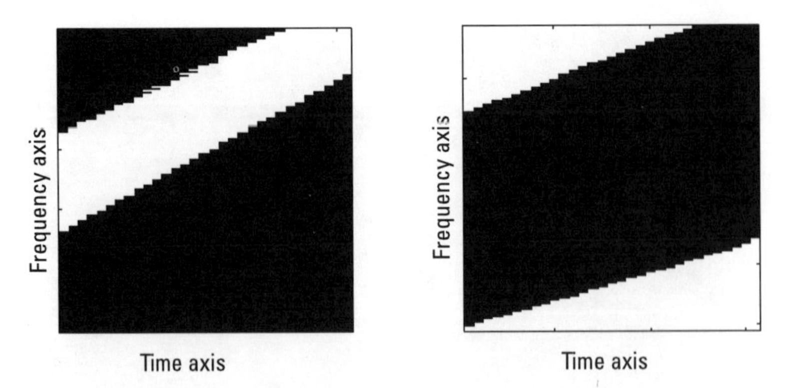


**Figure 10.18** Time-varying filters for selective motion compensation.

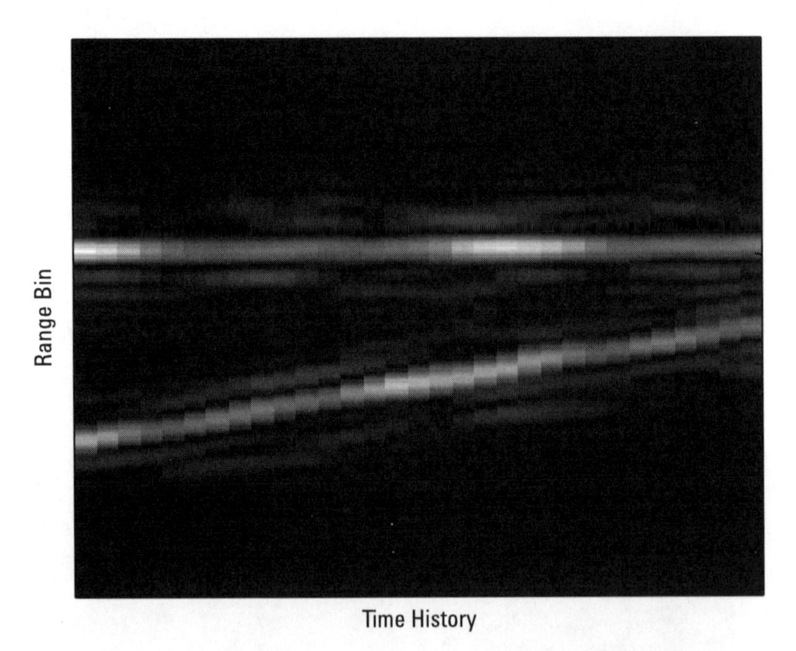


**Figure 10.19** Range profiles of the two aircraft without selective compensation.

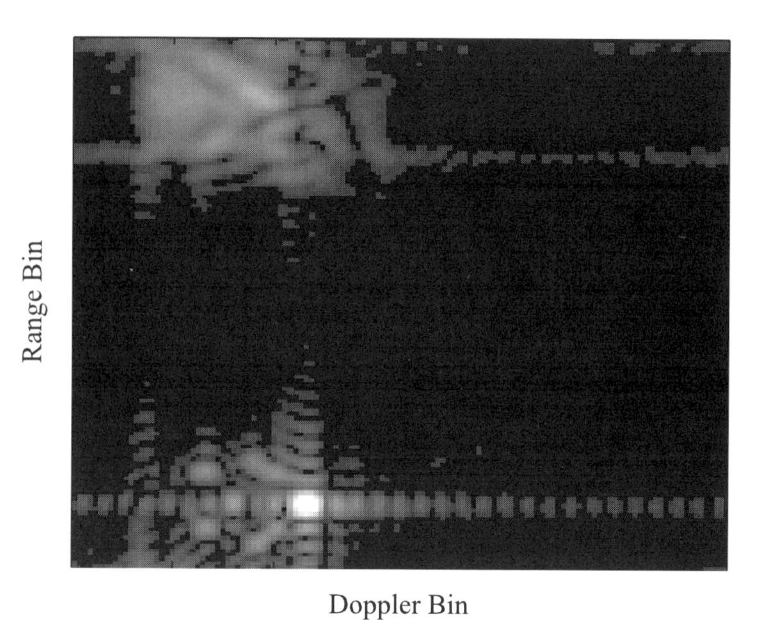


**Figure 10.20** ISAR image without selective motion compensation.

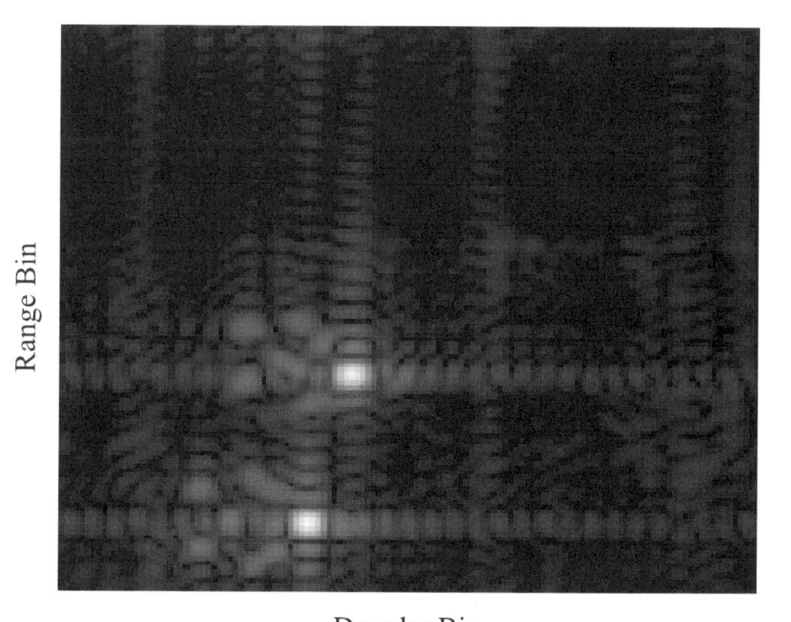


**Figure 10.21** Final image obtained using selective motion compensation.

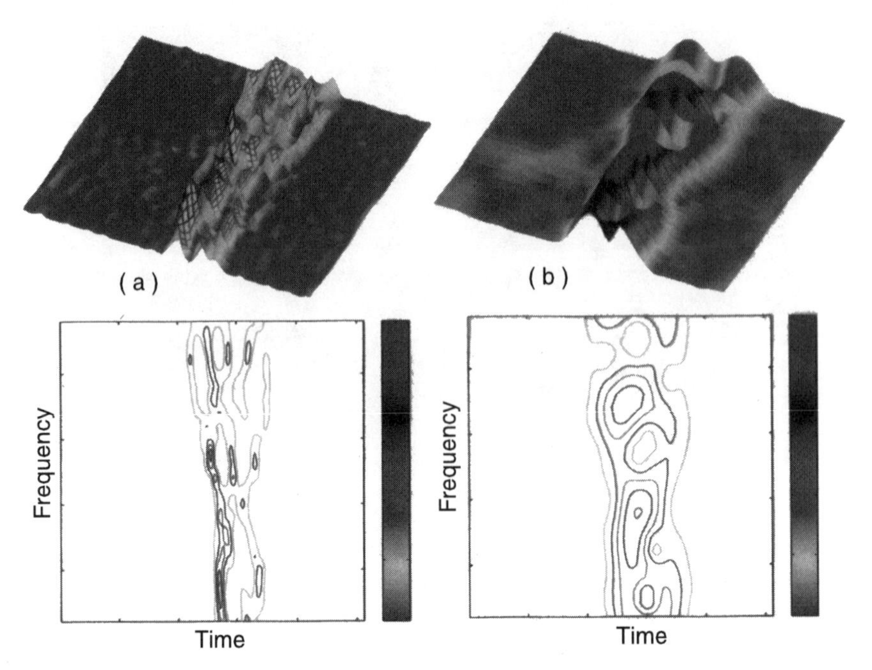


**Figure 10.22** Analysis of ISAR data using (a) a high-resolution TF transform and (b) the STFT.

## 10.7 Summary

The proposed method can successfully separate signatures of moving targets so that compensated images of each target can be obtained. CFAR filtering can be done either by using a time-varying approach based on the STFT, or by processing each burst. The filtering does not affect the estimation of the initial velocity and acceleration of the target, so focused images are obtained with those estimates. When signatures do not cross each other in the TF domain, blurring in each target's image can be effectively eliminated.

## References


[1] Kozek, W., and F. Hlawatsh, "A Comparative Study of Linear and Non-Linear Time-Frequency Filters," *Proc. of the IEEE-SP Int. Symp. On Time-Frequency and Time Scale Analysis*, Victoria, BC, Canada, Oct. 1992, pp. 163–166.

[2] Thomas, G., et al., "Evaluation of Time-Frequency Filtering for SAR/ISAR Motion Compensation via Instantaneous Frequency Estimators," *SPIE Radar Processing, Technology, and Applications*, Denver, Aug. 1996, Vol. 2845, pp. 141–150.

[3] Boashash, B., "Estimating and Interpreting the Instantaneous Frequency of a Signal—Part 2: Algorithms and Applications," *Proc. of the IEEE*, Vol. 80, No. 4, Apr. 1992, pp. 540–568.

[4] Kay, S. "Statistically/Computationally Efficient Frequency Estimation," *IEEE Proc. ICASSP88*, New York, Apr. 1988, Vol. W, pp. 2292–2295.

[5] Clarkson, P. M., *Optimal and Adaptive Signal Processing*, Boca Raton, FL: CRC Press, 1993.

[6] Haykin, S., *Adaptive Filter Theory*, 2nd ed., Englewood Cliffs, NJ: Prentice Hall, 1991.

[7] Boashash, B., "Time-Frequency Signal Analysis," in S. Haykin (ed.), *Advances in Spectrum Analysis and Array Processing*, Englewood Cliffs, NJ: Prentice Hall, 1991, pp. 418–517.

[8] Wong, K. M., and Q. Jin, "Estimation of the Time-Varying Frequency of a Signal: The Cramer-Rao Bound and the Application of Wigner Distribution," *IEEE Trans. on Signal Processing*, Vol. 38, No. 3, Mar. 1990, pp. 519–536.

[9] Boashash, B., and L. B. White, "Instantaneous Frequency Estimation and Automatic Time-Varying Filtering," *Proc. IEEE ICASSP9O*, Albuquerque, 1990, pp. 1221–1224.

[10] Whalen, A. D., and R. N McDonough, *Detection of Signals in Noise*, London: Academic Press, 1995.

[11] Giardina, C. R., and E. R. Dougherty, *Morphological Methods in Image and Signal Processing*, Englewood Cliffs, NJ: Prentice Hall, 1988.

[12] Stimson, G. W., *Introduction to Airborne Radar*, 2nd ed., Mendham, NJ: Science Publishing, 1998.

[13] Davies, E. R., *Machine Vision, Theory, Algorithms, Practicalities*, 2nd ed., Academic Press, 1997.

# 11

# Imaging Enhancement Using the Gabor Wavelet Transform

The use of the Gabor wavelet transform (GWT) has been proposed elsewhere [1] as a means to eliminate motion compensation. This chapter argues that, although motion compensation might be avoided when imaging a single target, this certainly does not apply for multiple moving target scenarios. Thus, the selective motion compensation methodology proposed in Chapter 10 is incorporated to the Gabor wavelet transform processing to obtain focused images.

## 11.1 Problem Statement

ISAR images of moving targets are obtained by processing doppler information on the content of a finite set of range cells. Doppler information (due to rotational or translational motion) typically is obtained using a doppler filterbank or a Fourier processor. The combination of target motion with long integration times degrades target images by shifting the target-scattering centers through different slant-range cells.

As discussed by Rihaczek [2], several images can be obtained using a short integration time. According to Chen [1], sufficient image resolution can be achieved by using the GWT. When the GWT parameters are properly chosen, the images obtained using that time-frequency representation possess higher quality than those obtained using long integration intervals if no motion compensation is performed. That is, image focus can be significantly improved by utilizating of shorter integration times.

This chapter discusses the selection of Gabor wavelet parameters. In particular, an analytical method for choosing the optimal integration time is presented. It should be noted that for images possessing multiple targets, the GWT parameters may differ for each target. In this instance, image segmentation by means of time-varying filtering can be incorporated in the GWT approach, thus providing focused images by reducing target phase errors separately.

## 11.2 Matched Filter Processing

To resolve target scatterers in cross-range, the doppler frequency associated with each scatterer needs to be determined. A doppler filterbank may be utilized to process the received signal. That approach maximizes the output SNR and optimizes the doppler parameter estimation [3]. Each filter in the bank is a narrowband network with impulse response

$$h(t) = s^*(t)e^{j2\pi f_D t} \tag{11.1}$$

where $s(t)$ is a replica of the transmitted signal and $f_D$ is the doppler frequency.

In practice, the doppler associated with the $k$th filter is

$$f_D = k\Delta f_D \tag{11.2}$$

where

$$\Delta f_D = \frac{1}{T} \tag{11.3}$$

and $T$ represents the signal duration.

The bandwidth of the doppler filterbank is determined by anticipating the bandwidth that characterizes the target [4]. Usually, the filter bandwidth is less than the signal's PRF.

For a point target, the output of the filterbank can be viewed as a series of slices of the signal's ambiguity function. The ambiguity function extends the idea of matched filtering. Surface concentrations in the ambiguity function plane determines the range-doppler positions of scatterers. The ambiguity function is based on the autocorrelation function of the received signal and is expressed as [5]

$$\chi(\tau, f_D) = \int_{-\infty}^{\infty} s(t)s^*(t-\tau)e^{j2\pi f_D t dt} \tag{11.4}$$

## 11.3 Coherent Processing and Image Formation

An alternative approach to extract doppler information is to coherently integrate successive samples of the signal at a given slant range. This is performed by using the Fourier transform. For better computational performance, the FFT algorithm is commonly used. Linking the Fourier transfer approach to a motion compensation procedure imagery with a high degree of doppler resolution. In practice, the Fourier transform approach is better than its filterbank counterpart because it allows for considerably longer integration times.

Target translational motion introduces undesirable range migration and doppler chirping effects, which are reflected in the exponential term outside the double integral in (2.1). Translational velocity and acceleration appear in that term as in (2.5). Those effects, in combination with similar effects introduced by rotational motion, cause significant blurring in the final image. Translational motion compensation algorithms estimate motion parameters and eliminate most of the phase errors introduced by the exponential term [6]. Coherent processing in $f_y$ is necessary to extract information in $y$ (i.e. cross-range). In practice, that is achieved by sampling the target signature in slow time at a fixed value of $x$. Fourier processing of that information is completed to extract the doppler target's components.

## 11.4 Effects of Target Motion

### 11.4.1 Range Walk and Range Offset

Target translational motion produces two detrimental effects: range walk between successive profiles and range offset within a single profile [5]. To show those distortions in an image, a single point scatterer is simulated according to (2.1), where a stepped frequency radar system is considered. In this example, the radar and scatterer parameters are as follows:

| | |
|---|---|
| $n = 64$ | Number of pulses |
| $N = 256$ | Number of bursts |
| $T_2 = 1.856E - 4$ | Pulse repetition interval (seconds) |
| $f_o = 3$ | Initial frequency (gigahertz) |
| $a_t = 0$ | Translational acceleration (meters per second squared) |
| $r = 6.5$ | Radius (meters) |
| $w$ | Angular velocity (radians per second) |
| $v_t$ | Translational velocity (meters per second) |

Range offset merely produces a fixed shift in range. Range walk produces blurring in the slant and cross-range direction, as shown in Figure 11.1(b). Figure 11.1(a) shows an image for zero translational velocity and short integration time. For a rapidly changing aspect angle, the effect of range migration is equivalent to reducing the integration time per range cell due to range displacement, as can be seen in Figure 11.1(c). At high velocities, the doppler information also spreads over a frequency band causing additional blurring, as shown in Figure 11.1(d).

### 11.4.2 Optimum Integration Time

The number of shifted range cells produced by range offset and range walk effects due to target motion can be approximately calculated for large values of $N$ by [5]

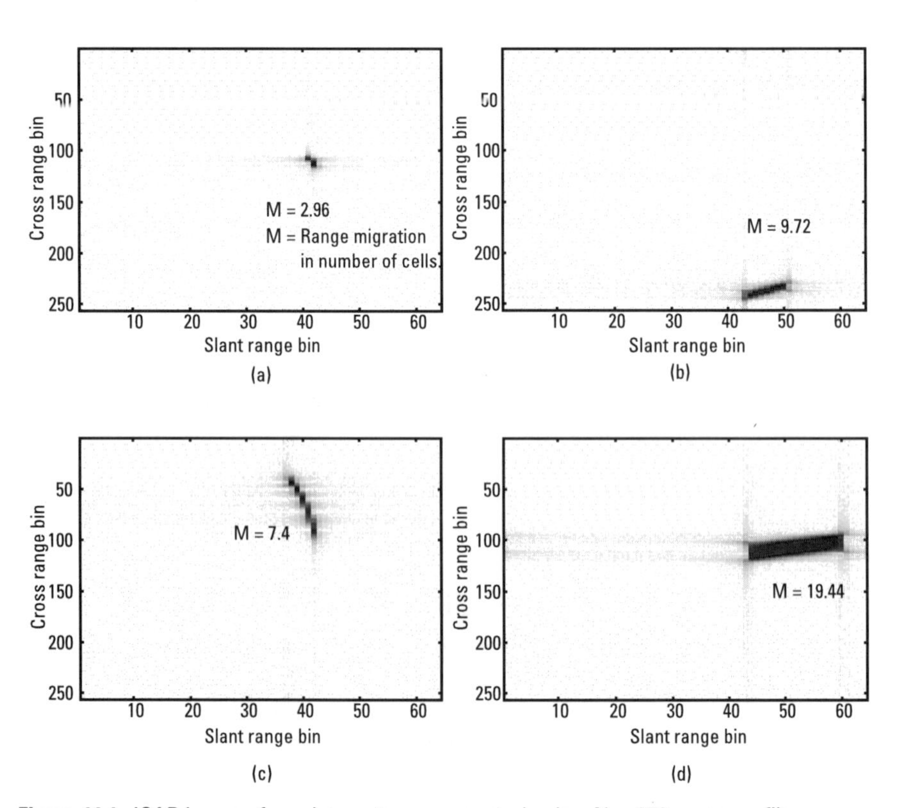


**Figure 11.1** ISAR image of a point scatterer generated using $N = 256$ range profiles.

$$M \approx \frac{2nT_2}{c}(f_0 + nN\Delta f)v_t \tag{11.5}$$

where $\Delta f$ denotes the frequency step between consecutive pulses.

For a rotating target, range migration is due to the radial movement of the scatterers. Likewise, image distortion increases as the integration time increases. The number of shifted range cells can be computed by

$$M = \frac{2\beta}{c}(wT_{\text{int}}) \tag{11.6}$$

where $T_{int} = nNT_2$ and $\beta = n\Delta f$. In Figure 11.1, migrations due to translational and rotational motion are calculated according to (11.5) and (11.6).

Nonlinear target maneuvers make it difficult to perform adequate motion compensation [7]. Using (11.5) and (11.6), however, it is possible to obtain the best integration time so that $M = 1$ and range migration does not occur in an image frame. Given $N$ bursts of data and a range migrate of $M$ cells, a set of images can be formed using the following number of bursts per image frame:

$$N' = \left[\frac{N}{M}\right] \tag{11.7}$$

The use of an $N'$ burst subaperture guarantees that each image obtained will have minimum range walk and range offset distortion due to target motion. The range profiles using the point scatterer example from the previous section and depicted in Figure 11.2 show that the maximum number of shifted cells may not exceed two, which is acceptable in terms of the image blurring it may cause. Optimum integration time should yield a range migration of one or, at worst, two cells.

Figure 11.3(a) shows an ISAR image of a simulated aircraft obtained using basic Fourier analysis over a long integration time. The blurring of the image is quite evident. Figure 11.3(b) shows that the optimum integration time yields better results even if it is shorter than the time used in Figure 11.3(a). Because range resolution is high, we can extract accurate feature information.

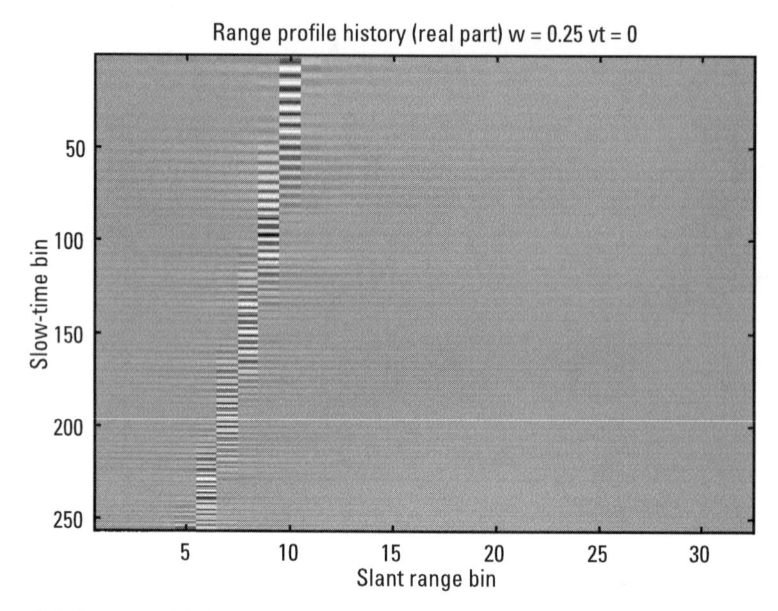


**Figure 11.2** Range profile history of a single point scatterer.

# 11.5 Applications of the GWT to ISAR Imaging

## 11.5.1 Selection of Gabor Function Parameters

The use of subapertures and integration times that optimize the formation of multiple ISAR images can be viewed as a time-frequency formulation of ISAR imaging. As proposed by Chen [1], the GWT can be used effectively for that purpose. In this case, the mother wavelet is defined by

$$G_p(t) = \frac{1}{\sqrt{2\pi\alpha_p^2}} \exp\left[-\frac{(t-\tau_p)}{2\alpha_p^2}\right] \exp[j2\pi f_p(t-\tau_p)] \qquad (11.8)$$

That function possesses the best tradeoff between time and frequency resolutions because the uncertainty product between those resolutions, $\Delta t \Delta f$, reaches the lower bound of 1/2. The set of parameters $(\alpha_p, \tau_p, f_p)$ defines a set of basis functions $g_p(t)$. The inner products formed by that set and the target's range profile history yield ISAR images [7].

According to the previous discussion, the shifting factor $\tau_p$ can be chosen so that range migration is equal to one or, at worst, two cells (refer to Figure 11.3). The Gaussian window modulation parameter $f_p$ in (11.8), is set to be

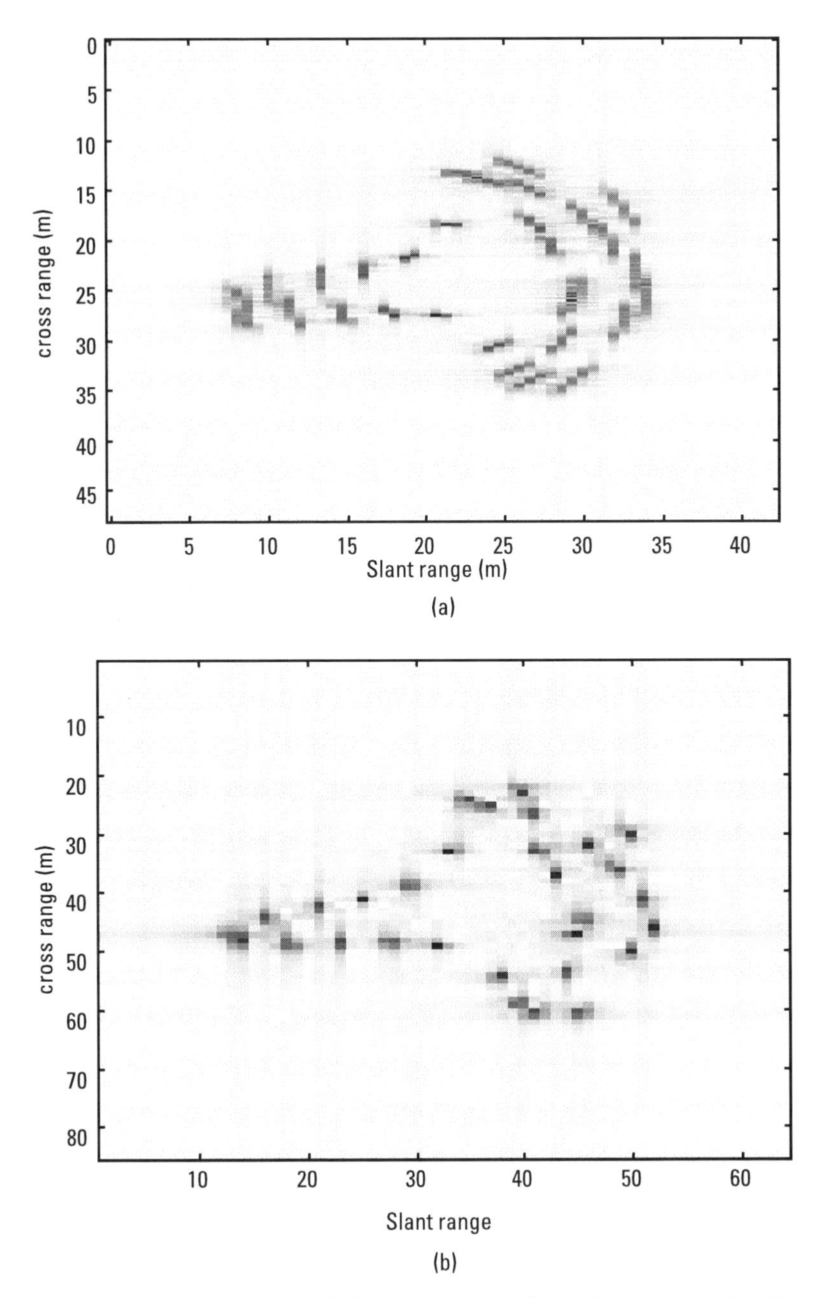


**Figure 11.3** ISAR images of simulated aircraft: (a) image of a moving target obtained by processing 256 profiles; (b) image obtained by processing 85 range profiles.

less than or equal to the PRF and expands around a preselected frequency band. The integration time of the ISAR frames is defined by $\alpha_p$. Note that that parameter corresponds to the standard deviation of the Gaussian window. Knowing that the area of a Gaussian function covered by twice the standard deviation is approximately 95%, we can safely set $\alpha_p$ as

$$2\alpha_p = N' \tag{11.9}$$

Following the preceding criteria and using the same parameters as the point scatterer example (with $w = 0.15$ rad/s and $vt = 2$ m/s), the optimum standard deviation value in this case is $\alpha_p = 13.1282$. The resulting image generated via the GWT approach is shown in Figure 11.4(a). To further illustrate the effect of $\alpha_p$ on the imaging process, other possible values were selected. Figure 11.4(b) was obtained by using a narrower window, whereas Figure 11.4(c) was obtained using a wider window. The optimum value for $\alpha_p$ yields the best image with minimum blurring.

### 11.5.2 Imaging of Multiple Moving Targets

If an image captures multiple targets moving at different velocities, the optimum $\alpha_p$ value will differ for each target. By choosing a particular value for $\alpha_p$, one target is focused. For the second target, the range offset differs for different frames, and the doppler frequency is still changing with time. That can be seen in Figure 11.5(b) and (c), where two images of simulated aircraft experiencing different velocities are shown. One target appears to be completely focused, while the other, though enhanced by the shorter integration time, does not exhibit the same resolution. Figure 11.5(a) shows an image in which the targets experience no translational motion, range migration is minimum, and longer integration times can be used. In Figure 11.5(b), both targets experience motion; therefore, a short integration time may not be optimum for both targets simultaneously. Figure 11.5(c) shows the consequence of using a long integration time, thus reducing resolution.

A possible approach to focusing both targets can be the separation of each target signature by means of time-frequency filtering. More specifically, the time history of a particular range cell can be processed using slow-time frequency filtering so that each aircraft can be separated and focused separately. Resolution using this approach may be poor, and consequently some objects may not be separated, as shown in Figures 11.7.

The problem is resolved by using the selective motion compensation approach (see Chapter 10), which utilizes all the available bursts. Because steps in

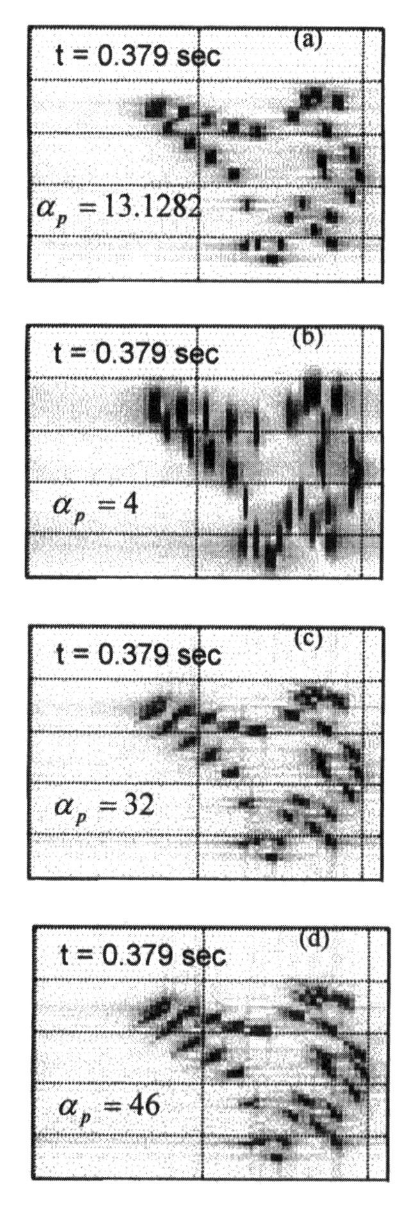


**Figure 11.4** Image frames generated via the GWT approach.

frequency appear in the exponential term in (2.1), bursts of data are filtered one at a time. That is similar to TF filtering via the STFT, in which the length of the analysis window is equal to the burst duration. To avoid discontinuities due to frequency stepping in (2.1), window overlapping is avoided during the

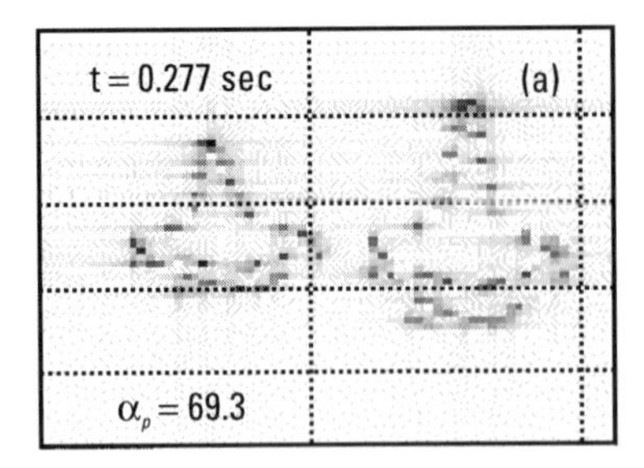


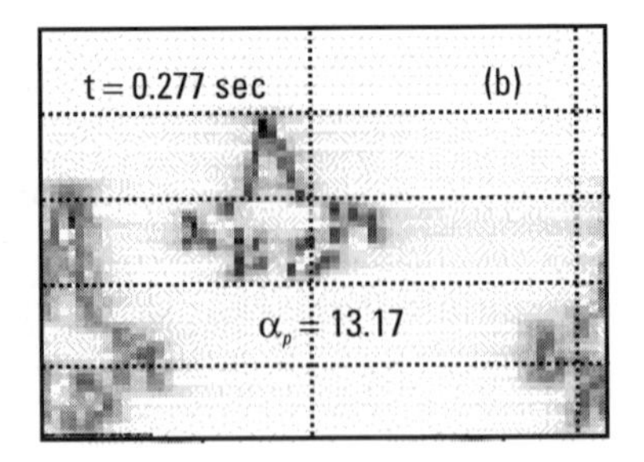


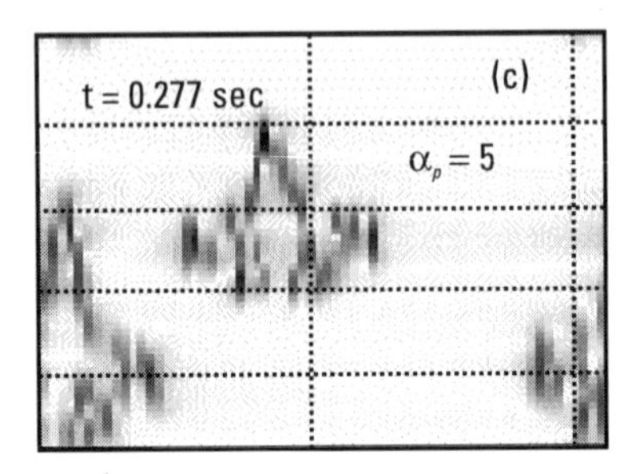


**Figure 11.5** Image frames from multiple targets generated via the GWT approach.

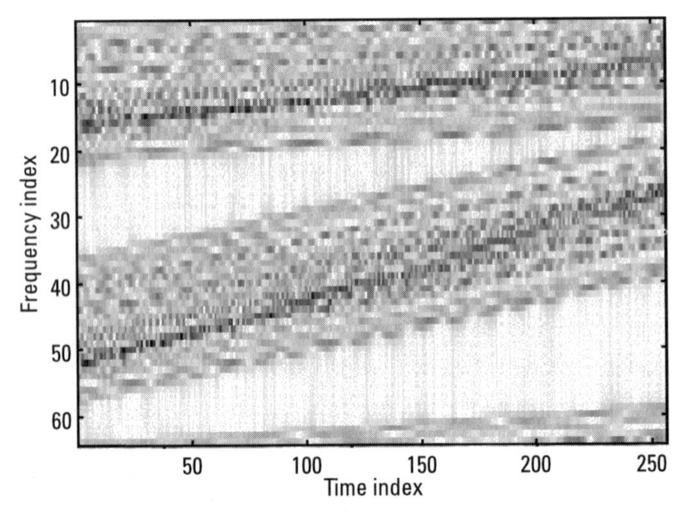


**Figure 11.6** STFT analysis using all available bursts. The two aircraft signatures can be separated.

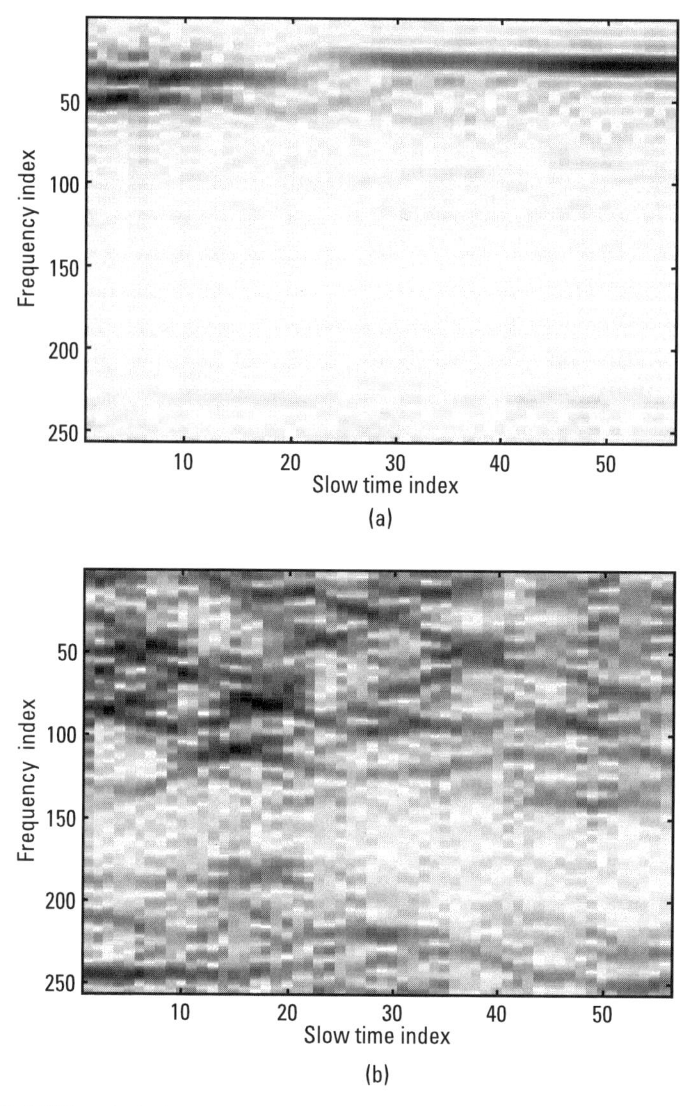


**Figure 11.7** STFT analysis for two fixed range cells.

analysis. To obtain the best possible reconstruction when the complete burst is used, a rectangular window is used. For example, the two aircraft signatures can be filtered by a TF mask, and compensation can be done separately, as shown in Figure 11.6. If an initial velocity estimate is computed by tracking the range

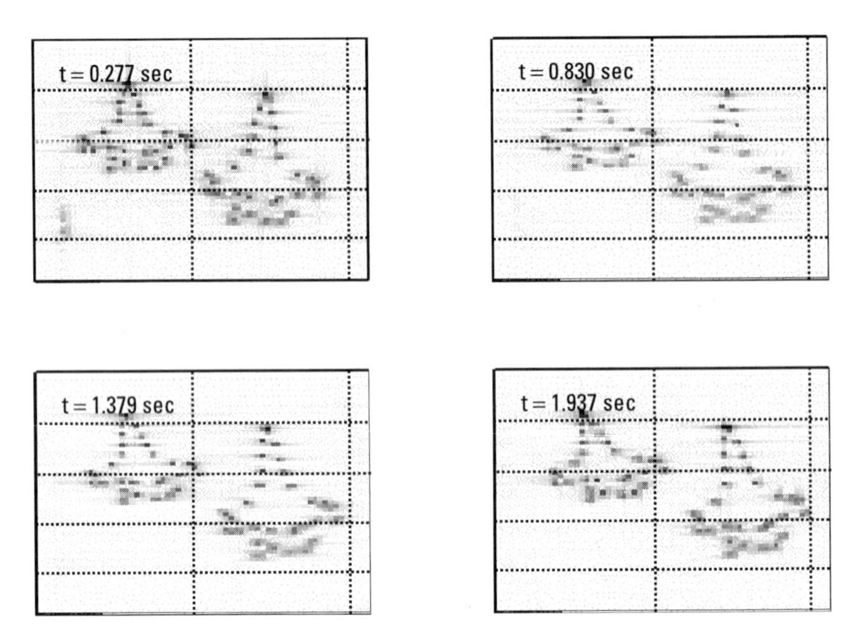


**Figure 11.8** ISAR frames using the GWT (standard deviation = 30) and selective motion compensation. Because phase errors have been reduced via motion compensation, images are highly focused.

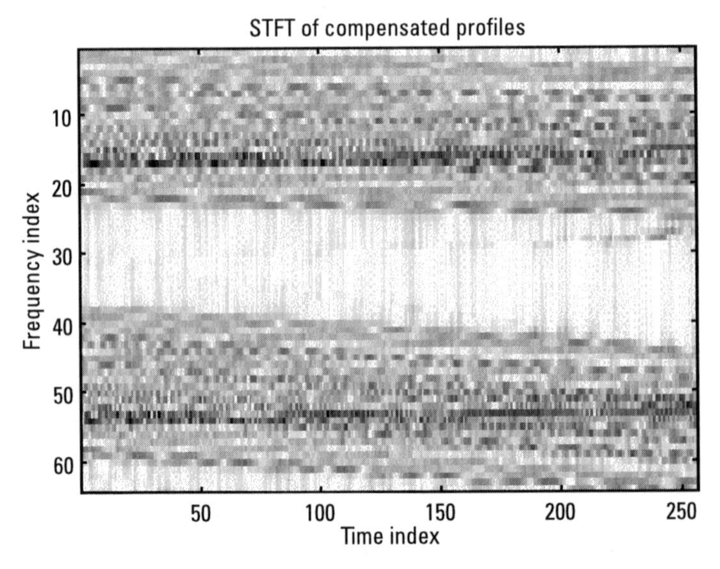


**Figure 11.9** Corrected profiles in which a velocity error of 20% was allowed. Compare with Figure 6.6(c).

profile peaks, phase errors can be corrected to a point where the GWT is effective. Results are shown in Figures 11.8 and 11.9, in which the velocity estimate correction has been performed and an error of 20% for each velocity estimation has been considered. The steps for obtaining the image are depicted in Figure 11.10.

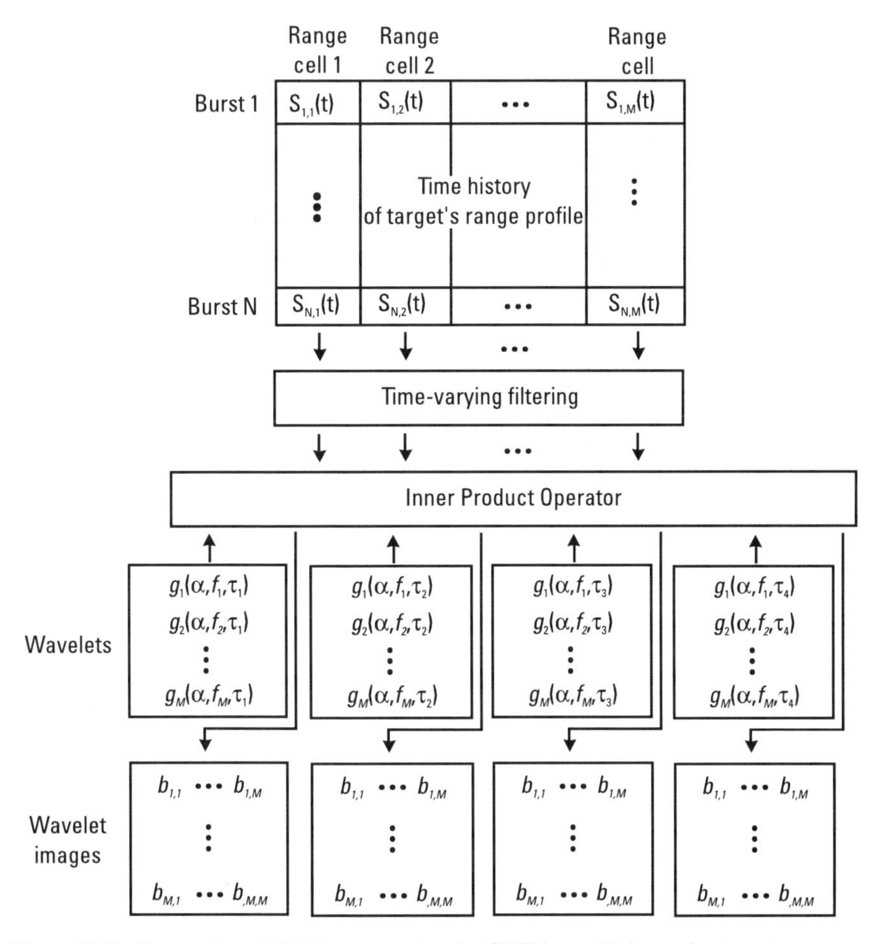


**Figure 11.10** Generation of ISAR images using the GWT for multiple moving targets.

# References

[1] Chen, V. C., "Reconstruction of Inverse Synthetic Aperture Radar Image Using Adaptive Time-Frequency Wavelet Transforms," *Proc. SPIE Wavelet Applications for Dual Use*, Orlando, FL, 1995, Vol. 2491, pp. 373–386.

[2] Rihaczek, A. W., and S. J. Hershkowitz, *Radar Resolution and Complex-Image Analysis*, Norwood, MA: Artech House, 1996.

[3] Skolnik, M. I., *Introduction to Radar Systems*, New York: McGraw-Hill, 1962.

[4] Stimson, G. W., *Introduction to Airborne Radar*, 2nd ed., Mendham, NJ: Science Publishing, 1998.

[5] Wehner, D. R., *High Resolution Radar*, 2nd ed., Norwood, MA: Artech House, 1995.

[6] Bocker, R. P., and S. A. Jones, "ISAR Motion Compensation Using the Burst Derivative Measure as a Focal Indicator," *Internat. J. of Imaging Systems and Technology*, Vol. 4, 1992, pp. 285–297.

[7] Son., J., B. C. Flores, and S. Tariq, "An Efficient Target Motion Compensation Method for Stepped Frequency ISAR Signatures," *Proc. SPIE Radar Processing, Technology, and Applications II*, San Diego, July 1997, Vol. 3161, pp. 20–28.

# 12

# Introduction to Rotational Motion Compensation

The essential requirement for ISAR imaging is target rotation. This type of motion produces the doppler shift required to map the target's reflectivity in the cross-range direction. However, to form an image using the two-dimensional DFT approach, target rotation must be small to avoid blurring of image features. Blurring or image defocusing is a combination of a varying cross-range and range walk. Time-varying doppler shifts associated with the individual target scattering points must be smaller than the doppler resolution to prevent distortion in cross-range. Similarly, the range shift incurred by the scatterers must be smaller than the down-range resolution to prevent distortion in the down range. Blurring is observed whenever the doppler shift of each scatterer deviates from a constant or if there is a rate of down-range change.

To form the focused ISAR image of a rotating target via the two-dimensional DFT, the array of samples $H(i, k)$ must be arranged in orthogonal directions in the frequency domain so that the samples may be transformed to an image plane formed by the down-range and cross-range directional vectors. Thus, the image generated can be perfectly focused if the angular motion parameters are known. When $H(i, k)$ is not arranged in a rectangular grid, the image will be focused only if the maximum dimension of the target projected onto the image plane is smaller than the blur radius. For a target whose dimensions are larger than the blur radius, the image quality will degrade as resolution is reduced and the wavelength is increased.

This chapter discusses a data interpolation technique in the frequency domain that allows us to retain the computational advantage of the two-dimensional DFT exploited in the imaging algorithm. In some experimental scenarios, data samples of the frequency response collected as the target rotates conform to a polar format where the data may be uniformly spaced in frequency and angular position [1, 2]. As such, the data samples do not match the rectangular format in the frequency plane needed for image processing. Resampling from polar format to rectangular format is done by interpolating the measured data in two dimensions orthogonal to each other. The resulting set of data, which is said to be polar reformatted, can be processed into a blur-free ISAR image.

## 12.1 Quadratic Phase Distortion

The doppler frequency of a rotating point scatterer at a distance $r$ from the center of rotation is expressed as

$$f_D = \frac{2f_c}{c}\omega r \cos(\omega t + \theta_0) \tag{12.1}$$

which is a time-dependent frequency shift. The target rotation angle during the image frame time $T$ is

$$\Delta\theta = (\omega T + \theta_0) - \theta_0 = \omega T \tag{12.2}$$

If the aspect change is significant, then the shift in doppler will force the spectrum to have a relatively wide main response, which leads to uncertainty in the cross-range position. The greater uncertainty leads to the greater blurring in the cross-range. Using (12.1), we find that doppler widening is related to the range uncertainty by

$$\delta f_D = \frac{2\omega f_c}{c}\delta r_c \tag{12.3}$$

## 12.2 Frequency Space Aperture

Figure 12.1 depicts a set of frequency responses collected in polar format. In this map, the radial distance to the origin represents the frequency. The LOS angle is represented by the angular position with respect to the horizontal axis.

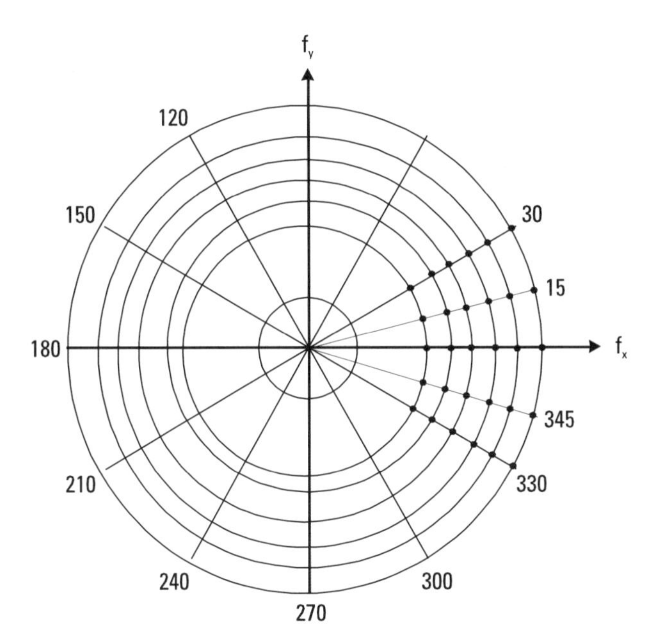


**Figure 12.1** Location of sampled points in the frequency space.

Although the frequency and angular spacings are uniform, the samples do not conform to the orthogonality of a cartesian plane. For this geometry, the rectangular coordinates for a data sample are given by [3, 4]

$$f_x = \frac{2f}{c}\cos(\theta) \tag{12.4}$$

and

$$f_y = \frac{2f}{c}\cos(\theta) \tag{12.5}$$

Then the baseband signal in the frequency domain is written in terms of target coordinates $(x, y)$ as the two-dimensional Fourier transform of the space-domain reflectivity:

$$F_2[g(x, y)] = \int_{-\infty}^{\infty}\int_{-\infty}^{\infty} g(x, y)\exp[-j2\pi(f_x x + f_x x)]dxdy \tag{12.6}$$

Here, $F_2$ is the two-dimensional Fourier transform operator. If we let $G(f_x, f_y) = F_2\{g(x, y)\}$, then the two-dimensional Inverse Fourier transform is given by

$$F_2^{-1}[G(f_x, f_y)] = \int_{-\infty}^{\infty}\int_{-\infty}^{\infty} G(f_x, f_y) \exp[j2\pi(f_x x + f_y y)] df_x df_y \quad (12.7)$$

Because measurements in ISAR usually are obtained in polar coordinates, it is appropriate to show the polar version of the Fourier transform in this domain. Hence, we let

$$x = \rho \cos \phi \quad \text{and} \quad y = \rho \sin \phi \quad (12.8)$$

where

$$\rho = \sqrt{x^2 + y^2} \quad (12.9)$$

Here, $\rho$ and $\phi$ are the polar coordinates in the image domain. Then, the Fourier transform of $g$ can be written as:

$$F_2[g(\rho, \phi)] = \int_0^{2\pi}\int_0^{\infty} g(\rho, \phi) \exp[-jr\rho \cos(\phi - \theta)]\rho d\rho d\phi \quad (12.10)$$

where

$$r = \sqrt{f_x^2 + f_y^2} \quad (12.11)$$

and

$$\theta = \tan^{-1}(f_y / f_x) \quad (12.12)$$

$f_x$ and $f_y$ are the spatial frequency coordinates in units of 1/m, and $\rho$ is the spatial radial frequency in units of 1/m.

## 12.3 Central Slice Projection Theorem

The slice projection theorem states that the Fourier transform of a projection of an ISAR target at an orientation $\phi$ is a "slice," or section, of the two-dimensional Fourier transform of the target's reflectivity, passing through the origin and oriented at the same angle [4].

Let $g(x, y)$ represent the object whose image is desired. When the radar beam illuminates the object, the output of the matched filter receiver represents the projection $g(x, y)$ onto the $u$ axis (i.e., the LOS). The projection at the given $\phi$ is a fraction of $u$ and it is represented by

$$p(u, \phi) = \int g(u \cos \phi - v \sin \phi, u \sin \phi + -v \cos \phi) dv \quad (12.13)$$

where

$$x = u \cos \phi - v \sin \phi \quad \text{and} \quad y = u \sin \phi + v \cos \phi \quad (12.14)$$

Mathematically, the projection of this object at the view angle $\phi$ is given by [4]

$$p_\phi(u) = \int_L g(u, v) dv \quad (12.15)$$

where $L$ is the cross-range integration path. The Fourier transform of $p_\phi(u)$ is given by

$$P_\phi(r) = \int_{-\infty}^{\infty} P_\phi(u) \exp(-j2\pi ur) du \quad (12.16)$$

Notice that $P_\phi(r)$ is the spatial frequency response of the object as determined for the viewing angle $\phi$ along the axis. Recall that the Fourier transform of the function $g(x, y)$ in polar coordinates is given by

$$G(u, \phi) = \int_{-\infty}^{\infty} \int_{-\infty}^{\infty} g(x, y) \exp[-j2\pi(ux \operatorname{cox} \phi + uy \sin \phi)] dxdy \quad (12.17)$$

Rewriting (12.17) in terms of the coordinate rotation relations, we have

$$G(r, \phi) = \int_{-\infty}^{\infty} \int_{-\infty}^{\infty} g(u \cos \phi - v \sin \theta, u \sin \phi + v \cos \theta) \exp(-j2\pi ur) dudv$$

$$(12.18)$$

which can be written in terms of projection as

$$G(r, \phi) = \int p(r, \phi) \exp(-j2\pi r u) \tag{12.19}$$

Equation (12.19) is equal to (12.16). Therefore,

$$G(r, \phi) = P(r, \phi) \tag{12.20}$$

The result in (12.20) states that the one-dimensional Fourier transform of the projection of an object at a given angle $\phi$ is a slice of the two-dimensional Fourier transform of the object oriented at the same angle.

In theory, if an infinite number of projections were available, a perfect reconstruction of the image of an object would be possible. In practice, however, the degree of resolution is limited by the collection of projections measured over a finite number of viewing angles and the length of such projections.

## 12.4 Blur Radius

The range extent of a target plays an important role in predicting in what regions of an image may be blurred. The radar signature of a target that is relatively small compared to the intended resolution does not require rotational motion compensation. The blur radius dictates a practical limit for target size beyond which the unfocused ISAR image becomes blurred at its edge. The blur radius is defined as the target radius that results in a maximum of one cell of slant-range or cross-range migration during the required integration angle to achieve a given cross-range resolution [3]. The target blur radius for one cell of migration for a square resolution ($\Delta r = \Delta r_c = \Delta r_s$) is expressed as

$$r = \frac{\Delta r}{\Delta \theta} = \frac{2(\Delta r)^2}{\lambda} \tag{12.21}$$

Figure 12.2(a) shows the image of a rotating target composed with multiple point scatterers. Four point scatterers are within the blur radius. Even though the target rotates, scatterers within the blur radius are focused; however, the scatterers outside the blur radius are not focused. Notice that the blurring effect is more prominent for a scatterer farther from the rotating center. Figure 12.2(b) shows the rotational motion compensated version of Figure 12.2(a). In this figure, the scatterers are focused everywhere.

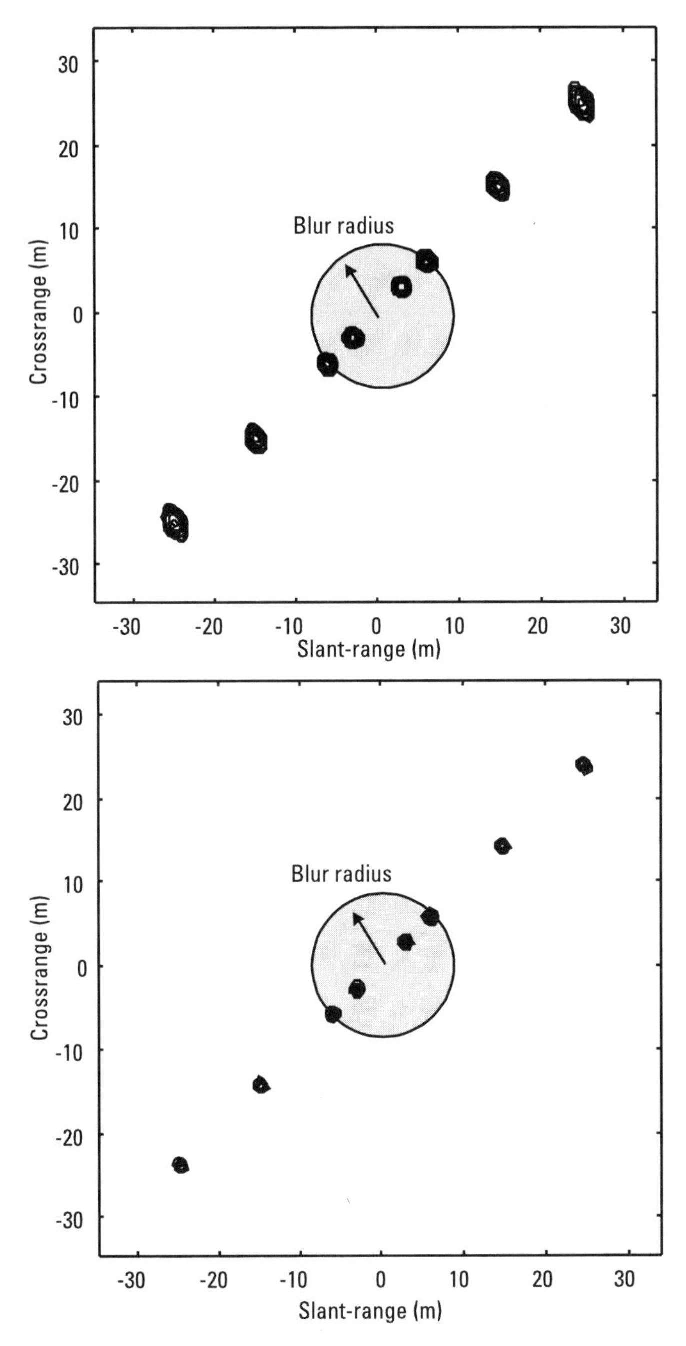


**Figure 12.2** (a) Unfocused image outside blur radius; (b) focused image after rotational motion compensation.

## 12.5 ISAR Geometry

One of the features of ISAR signatures is that they are collected over a band-limited region or annulus. This leads to the necessity of obtaining an image based on an incomplete set of data measurements. This capability is illustrated in Figure 12.1, which shows an annulus of polar coordinates representing the original band-limited sampled region. The interpolation process is performed over a rectangular band-limited region within the original polar region. The length and width of that region are given by (12.22). The parameters $\Delta f_x$ and $\Delta f_y$ are given by (12.23).

$$B_x = (N - 1)\Delta f_x = [f_{N-1} \cos(\Delta\theta/2) - f_o](2/c)$$

$$B_y = (M - 1)\Delta f_y = [2 f_0 \tan(\Delta\theta/2) - f_o](2/c) \tag{12.22}$$

and

$$\Delta f_x = (f_{x(N-1)} - f_{xo})/(n - 1)$$

$$\Delta f_y = \left[\frac{2 f_0}{M - 1} \tan(\Delta\theta/2)\right](2/c) \tag{12.23}$$

where $M$ is the number of bursts, $N$ is the number of pulses per burst, and $\Delta\theta$ is the aspect change of the target. The new coordinates after resampling will be given in terms of $\Delta f_x$ and $\Delta f_y$.

## 12.6 Slant-Range and Cross-Range Resolution

To produce an undistorted image via the two-dimensional DFT, the data must be interpolated to produce samples at the intersections of a rectangular grid. The new grid has uniform spaces defined as $\Delta f_x$ and $\Delta f_y$, which determine the length of the windows in the down-range and cross-range, respectively [3, 4]. Similarly, the dimensions of the grid determine the resolution in these ranges. For an $M$-by-$N$) grid, the extent of the ISAR image is given by

$$W_s = \frac{t}{\ell \Delta f_x} \tag{12.24}$$

in the down range and

$$W_c = \frac{1}{\Delta f_y} \tag{12.25}$$

in cross-range. Hence, down-range resolution is

$$\Delta r_s = \frac{1}{(N-1)\Delta f_x} \tag{12.26}$$

whereas the cross-range resolution is

$$\Delta r_c = \frac{1}{(M-1)\Delta f_y} \tag{12.27}$$

Through polar reformatting, the image size is increased in proportion to the aspect change. For a fixed number of grid samples, resolution in both dimensions deteriorates as the aspect change increases. Conventional weighting will further degrade the resolution of the ISAR image. Overall, image distortion is eliminated at the expense of resolution.

## 12.7 Stepped Rotation

In some reflectivity measurements, target rotation is purposely controlled to characterize the target at multiple viewing angles [4–6]. The target, which is to be mapped in down range and azimuth at microwave frequencies, is usually placed on a motor-driven rotary platform (or turntable). The aspect presented to the radar receiver is then changed discretely in $N$ small steps $\delta\theta$ over an angular interval $\Delta\theta$. Down-range profiles are obtained in the usual manner, using frequency-modulated or frequency-stepped waveforms. Cross-range profiles, however, cannot be obtained from doppler measurements because motion is no longer time dependent. Rather, cross-range information is obtained by analysis of phase changes as a function of aspect [7]. The gradient of the frequency response phase is

$$\frac{d\phi}{d\theta} = \frac{4\pi f}{c} r\cos(\theta) = \frac{4\pi f}{c} r_c \tag{12.28}$$

where $\theta$ is the aspect of the target, and $r_c$ is the cross-range position. Thus, for fixed frequencies, the phase gradient and cross-range are directly proportional to each other. Furthermore, because the phase gradient is proportional to the gradient of $f_y$, to identify scatterers in the cross-range, we need only transform data from the $f_y$ domain to the cross-range domain. Following that reasoning, we conclude that cross-range resolution is achieved by measuring and processing phase changes, not time-dependent doppler shifts.

The same aspect change that provides resolution in cross-range can complicate the processing of the ISAR image. For stepped-aspect changes, polar reformatting is performed in two separate one-dimensional steps as follows. First, a down-range interpolation is done by resampling I/Q data in the radial direction (frequency) while keeping $\theta$ fixed. Second, an azimuth interpolation is implemented on the intermediate I/Q sample grid from the previous step. Here, interpolation is performed along vertical parallel lines to produce the final rectangular grid. When the two-dimensional interpolation is completed, the data fits into a rectangular grid equally spaced in $f_x$ and $f_y$. The data are then converted into an ISAR image using the standard two-dimensional DFT.

### 12.7.1 Down-Range Interpolation

In this step, the new samples are to fall on positions along the vertical lines that form the rectangular grid. Line separation is

$$\Delta f_x = \frac{f_{x(N-1)} - f_{x0}}{N - 1} \tag{12.29}$$

where

$$f_{x(N-1)} = \frac{2}{c} f_{N-1} \cos(\Delta\theta/2) \tag{12.30}$$

and

$$f_{x0} = \frac{2}{c} f_0 \tag{12.31}$$

In (12.30) and (12.31), $f_0$ and $f_{N-1}$ are lower and upper synthetic frequencies, respectively. The resulting samples will he spaced according to

$$\delta f_x = \frac{(f_{x0} + \beta) \cos(\Delta\theta/2) - f_{x0}}{} \tag{12.32}$$

From Figure 5.3, the new polar coordinates used for the first interpolation step are

$$\rho_{ik} = \frac{f_{x(i)}}{\cos(\theta_k)} \tag{12.33}$$

for $i = 0, \ldots, N - 1$, and $k = 0, \ldots, M - 1$. In (12.12) the spatial frequency $f_x$ is stepped in equal increments:

$$f_{x(i)} = i\Delta f_x - f_{x0} \tag{12.34}$$

That process can be constructed as a digital filtering technique in which the input is a uniformly spaced sequence and the output is a sequence with a lower sampling rate and some sample delay. Sampling rate and sample delay are specified by the aspect change and viewing angle. When the number of interpolated samples is less than the number of input samples, as in the case of oversampled data, low-pass filtering of the original data must be performed to avoid aliasing effects in the down range [4].

### 12.7.2 Azimuth Interpolation

This step interpolates samples along parallel $f_x$ lines to create a uniform distribution in $f_y$, as shown in Figure 12.3. Thus, resampling is to be done at equal intervals $\Delta f_y$ given by

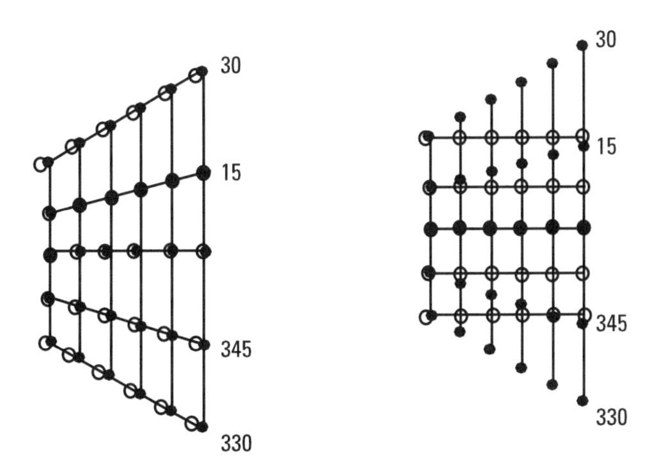


**Figure 12.3** Steps of interpolation for a target with stepped rotation.

$$\Delta f_y = \frac{2f_{x0}}{M-1}\tan(\Delta\theta/2) \tag{12.35}$$

For a fixed $f_{x(i)}$, the ordinate of the samples prior to azimuth interpolation is

$$f_{y(k)} = f_{x(i)}\tan(\theta_k) \tag{12.36}$$

On resampling, the ordinates of the data are given by

$$f_{y(k)} = \left(k - \frac{M-1}{2}\right)\Delta f_y \tag{12.37}$$

for $k = 0, \ldots, M - 1$. The resulting samples will be spaced according to

$$\delta f_y = \frac{f_0 \tan(\Delta\theta/2)}{M-1} \tag{12.38}$$

Interpolation in azimuth can also be viewed as a filtering operation in which a column vector of unequally spaced input samples produces a column vector of equally spaced output samples. When the original data are oversampled, a low-pass filter must be used to reduce the bandwidth of the sequence. The purpose of the filter is to match the bandwidth associated with the extent of the new cross-range window.

## 12.8 Continuous Rotation

Real targets such as ships and aircraft do not go through discrete angular motion in their natural environment. Relative to the radar, rotational motion is better approximated by an expression involving instantaneous parameters. For example, angular motion can be modeled as

$$\theta(t) = \theta_0 + w_0 t + \frac{1}{2}\alpha t^2 \tag{12.39}$$

where $\theta(t)$ is the target's instantaneous viewing angle at time $t$, $\theta_0$ is the initial angular position, $w$ is the initial angular rate, and $\alpha$ is the angular acceleration. Significant target migration may be expected while the stepped-frequency data is collected. Figure 12.4 illustrates the frequency response gathered as the target rotates at a constant angular rate $w$ (i.e., set $\alpha = 0$ rad/s$^2$).

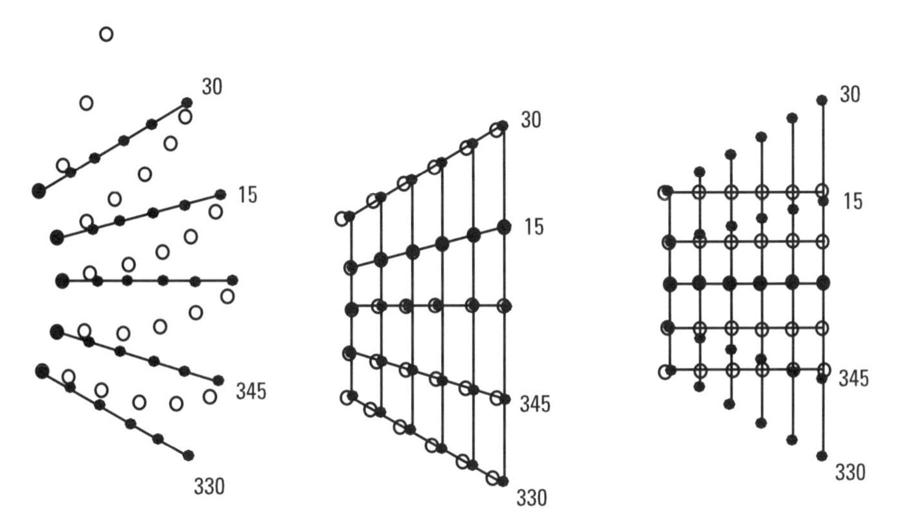


**Figure 12.4** Steps for interpolation for a target with continuous rotation.

Before the reformatting process can be carried out in down range and azimuth, the curved data first must be resampled into discrete, uniformly spaced polar angles. This step involves interpolating equally spaced data over concentric arcs for each synthetic frequency. The desired polar angle is

$$\theta_k = \left(k - \frac{M-1}{2}\right)\delta\theta \tag{12.40}$$

where $\theta_k$ is the angular migration between bursts, for $k = 0, \ldots, M-1$. The angular position for each sample before interpolation is

$$\theta_{ik} = i\frac{\delta\theta}{N} + \theta_k \tag{12.41}$$

where $i = 0, \ldots, N-1$ and $N$ is the number of pulses per burst. Referring to the first grid in Figure 12.4, it can be seen that some extrapolation is required for $k = 0$. Extrapolation may introduce an error that increases for wider arcs, that is, higher frequencies. The error, however, is negligible if the frequency response varies slowly for small aspect steps $\delta\theta$.

## References

[1] Chen, C., and H. C. Andrews, "Target-Motion-Induced Radar Imaging," *IEEE Trans. on Aerospace and Electronics Systems*, Vol. AES-16, No. 1, Jan. 1980, pp. 2–14

[2] Chen, C., and H. C. Andrews, "Multifrequency Imaging of Radar Turntable Data," *IEEE Trans. on Aerospace and Electronics Systems*, Vol. AES-16, No. 1, Jan. 1980, pp. 15–22.

[3] Wehner, D. R., High Resolution Radar, Norwood, MA: Artech House, 1987.

[4] Mensa, D., *High Resolution Radar Cross Section Imaging*, Norwood, MA: Artech House, 1990.

[5] Ausherman, D. A., et al., "Developments in Radar Imaging," *IEEE Trans. on Aerospace and Electronics Systems*, Vol. AES-20, No. 4, July 1984, pp. 363–400.

[6] Walker , J. L., "Range-Doppler Imaging of Rotating Objects," *IEEE Trans. on Aerospace and Electronics Systems*, Vol. AES-16, No. 1, Jan. 1980, pp. 23–52.

[7] Meloling, J. H., "A Study of the Interactions Between Curved Surfaces," Master's thesis, Arizona State Univ., Aug. 1989.

# 13

# Interpolation Methods for Rotational Motion Compensation

Interpolation is a numerical analysis technique used in science and engineering to fit a function to a given set of data. The technique can be done in different ways. One-dimensional algorithms may employ as few as two points from a data set to assign a value to a point between concurrent data samples. Other, more ambitious methods use more data points for the same purpose. Some methods referred to as weighted integration interpolation techniques, consider all available available samples to calculate the value of an unknown point within one- or two-dimensional data grids.

All those techniques approximate the value of a function at a given point. Because only a finite number of original data samples are available when interpolation is used, a closed-form formula is seldom available for the sampled function. Therefore, the value of a point located between two or more samples has to be approximated within a margin of error.

Polar reformatting is an interpolation process whereby a set of samples uniformly spaced in rectangular coordinates is computed from another set of samples uniformly spaced in cylindrical coordinates. The rectangular grid formed by the coordinates of the interpolated samples is contained within the boundaries of the cylindrical grid corresponding to the coordinates of the original samples. The accuracy of polar reformatting depends on the technique selected and on the spacing between samples.

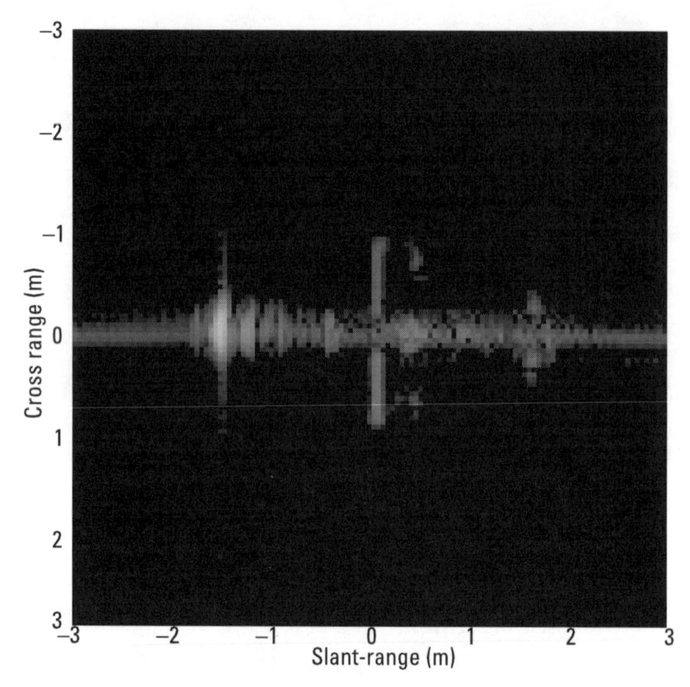


**Figure 13.1** Image for drone model obtained without rotational motion compensation. (Experimental data courtesy of Naval Air Warfare Center, Point Mugu, California.)

## 13.1 Approximation Methods

In the context of polar reformatting, approximation methods are ad hoc approaches that do not attempt to fit a particular function through the data. The nearest-neighbor and euclidean approximation techniques are the most frequently considered methods in this category [1, 2].

### 13.1.1 Nearest-Neighbor Approximation

In the nearest-neighbor approximation, an unknown point is assigned the value of the closest data point in a two-dimensional grid. For small aspect changes, the cylindrical grid is approximately rectangular, and any unknown point will fall inside a quasi-rectangular area (Figure 13.2) from which the distance to the closest point can be determined. Due to its simplistic nature, of the nearest-neighbor approach usually introduces large errors.

Figure 13.3 shows the result of rotational motion compensation using the nearest-neighbor approximation interpolation method. This figure confirms that significant errors are introduced using this approach.

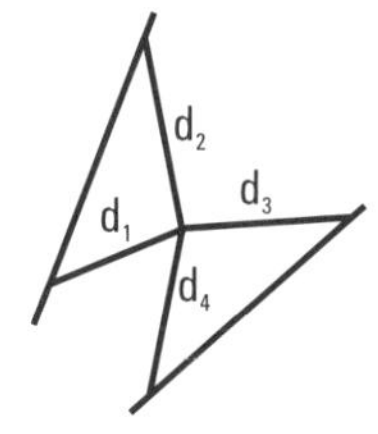


**Figure 13.2** Interpolation geometry using nearest-neighbor and euclidean approximations.

### 13.1.2 Euclidean Approximation

The Euclidean approximation technique uses a weighted average between the amplitudes of the four data samples surrounding the unknown point. For a polar grid, it is possible to define a cell as the space confined by the four samples, as illustrated in Figure 13.1. The value of an interpolated sample is calculated as follows [2]:

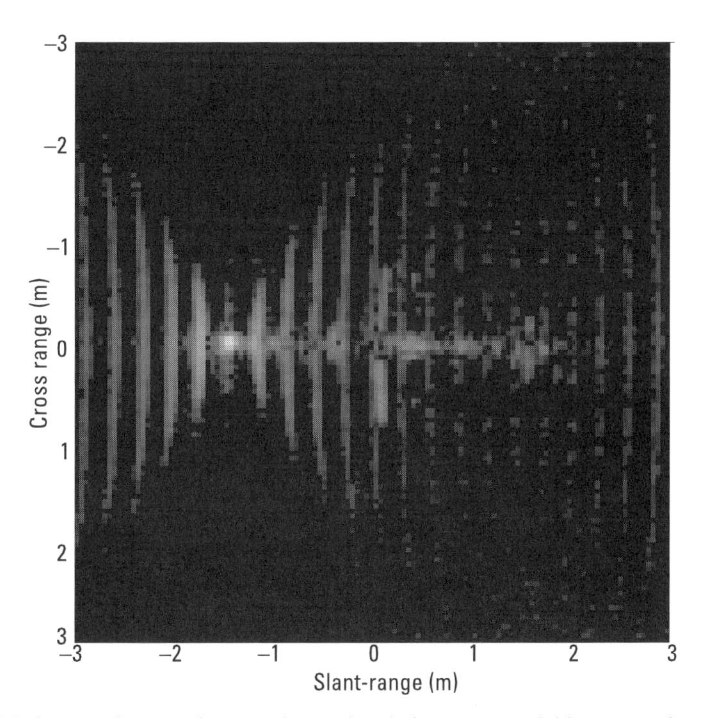


**Figure 13.3** Image of target signature interpolated via nearest neighbor approximation. (Experimental data courtesy of Naval Air Warfare Center, Point Mugu, California.)

$$f(x) = \frac{\sum_{m=1}^{4} f(x_m)/d_m}{\sum_{n=1}^{4} 1/d_n} \tag{13.1}$$

where $f(x_m)$ and $d_m$ are the amplitudes of the samples and their distances to the unknown point, respectively. The formula for $f(x)$ expressed in (13.1) depends on the actual distance between the sample points and the unknown point, ignoring any directional information (i.e., the coordinates of the data samples are used only to find the distance to the point in terms of magnitude). This may not be the best approach to approximate two-dimensional signals. The image obtained by the euclidean approximation interpolation method for rotational motion compensation is shown in Figure 13.4.

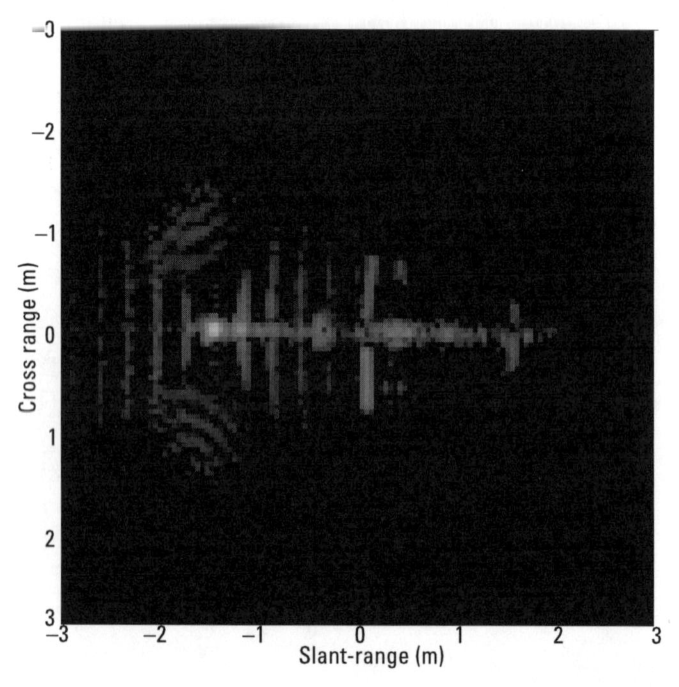


**Figure 13.4** Image of target signature interpolated via euclidean approximation. (Experimental data courtesy of Naval Air Warfare Center, Point Mugu, California.)

## 13.2 One-Dimensional Interpolation Methods

Some interpolation methods attempt to fit a function between two or more data samples, assigning a value to an unknown point based on the resulting "fitted" function. Two-dimensional functions can be interpolated through several one-dimensional interpolation steps. The three most popular one-dimensional functions used to interpolate are linear functions, cubic splines, and sinc functions. Figure 13.5 illustrates the different results those functions may yield when the number of samples is limited but no other restrictions apply. The amplitudes of the samples to be interpolated were chosen randomly.

### 13.2.1 Linear Interpolation

Interpolation with linear functions is known as first-order interpolation and is defined by the LaGrange formula for linear interpolation [3]

$$f(x) = [(x_1 - x)f_0 + (x - x_0)f_1]/(x_1 - x_0) \tag{13.2}$$

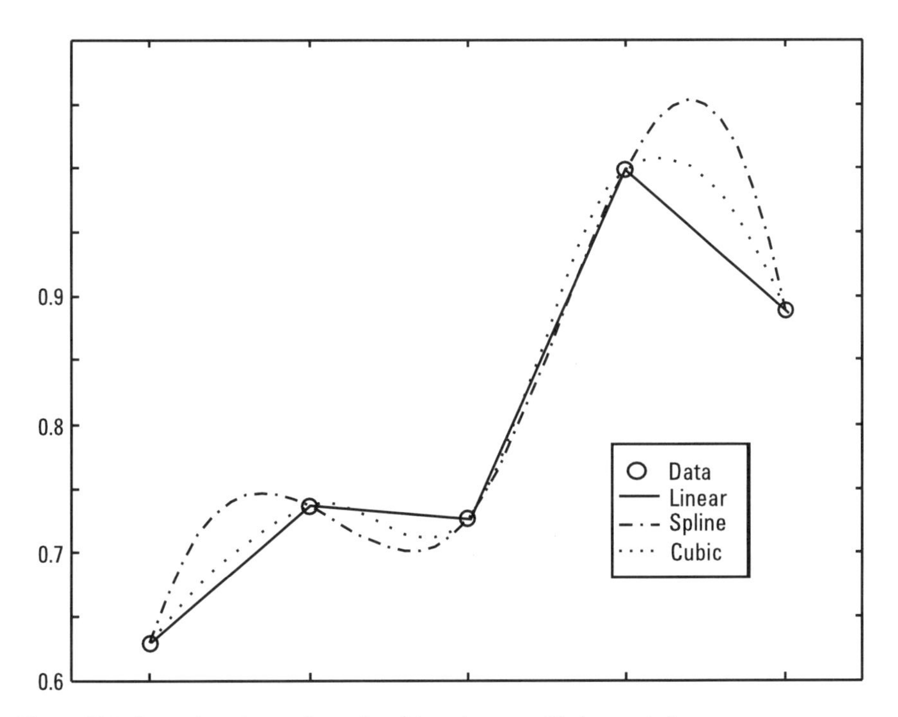


**Figure 13.5** Examples of one-dimensional functions used in interpolation.

where $(x_0, f_0)$ and $(x_1, f_1)$ are the known positions and values of two adjacent data samples. This simple approach usually requires sampling in the spatial frequency domain at a rate greater than the Nyquist rate to interpolate accurately between sampled values of the target's signature. Figure 13.6 shows the effects of using linear interpolation to compensate for rotational motion.

### 13.2.2 Cubic Spline Interpolation

Splines are segments of polynomial functions used as an interpolation basis to model a sampled function so that a set of $n + 1$ data points can be represented by $n$ splines. Cubic splines (i.e., third-order polynomials) typically are the ones most commonly used. Adjacent splines keep continuity in their curvature and slope at the knots, or extremes of the spline [4]. A cubic spline requires 4 coefficients. Thus, $4n$ coefficients are needed to interpolate a data set consisting of $n + 1$ data samples. Endpoint conditions are exploited to determine the coefficients for the first and last splines. One of the possible conditions to assume is the "not-a-knot" condition, which assumes that nothing is known about the

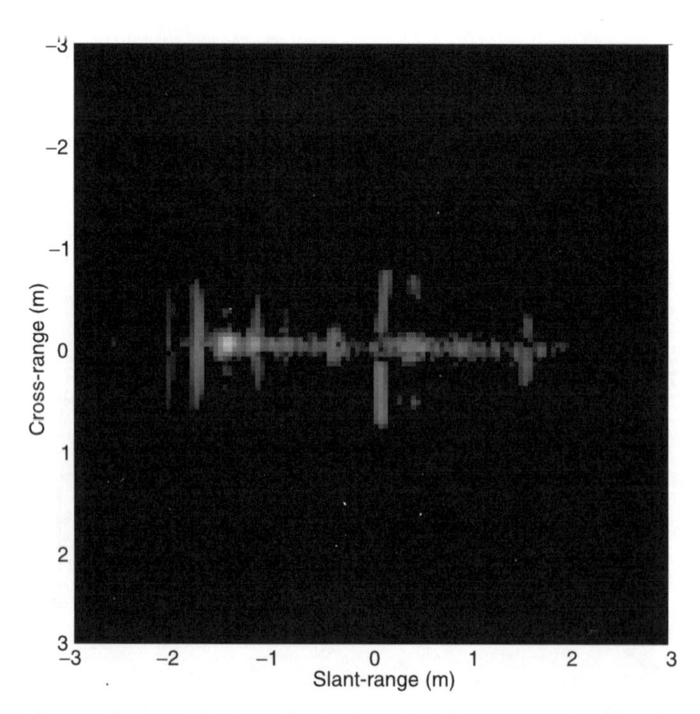


**Figure 13.6** Image of target signature interpolated via linear approximation. (Experimental data courtesy of Naval Air Warfare Center, Point Mugu, California.)

derivatives at the endpoints [3, 4]. Spline interpolation implemented on Matlab® uses that condition.

Splines with the not-a-knot condition approach a parabola at the ends (i.e. the second and second-to-last points are not active), as shown on Figure 13.7. As defined in [3], the $i$th cubic spline is given by

$$g_i(x) = a_i(x - x_i)^3 + b_i(x - x_i)^2 + c_i(x - x_i) + d_i \qquad (13.3)$$

where

$$\begin{aligned} a_i &= (S_{i+1} - S_i)/6h_i, \qquad b_i = S_i/2 \\ c_i &= [(y_{i+1} - y_i)/h_i] - (2h_iS_i + h_iS_{i+1})/6 \qquad d_i = y_I \end{aligned} \qquad (13.4)$$

and $S_i = g_i''(x_i)$, where $i = 0, 1, \ldots, n - 1$ and $S_n = g_{n-1}''(x_n)$. In matrix notation, the not-a-knot condition gives

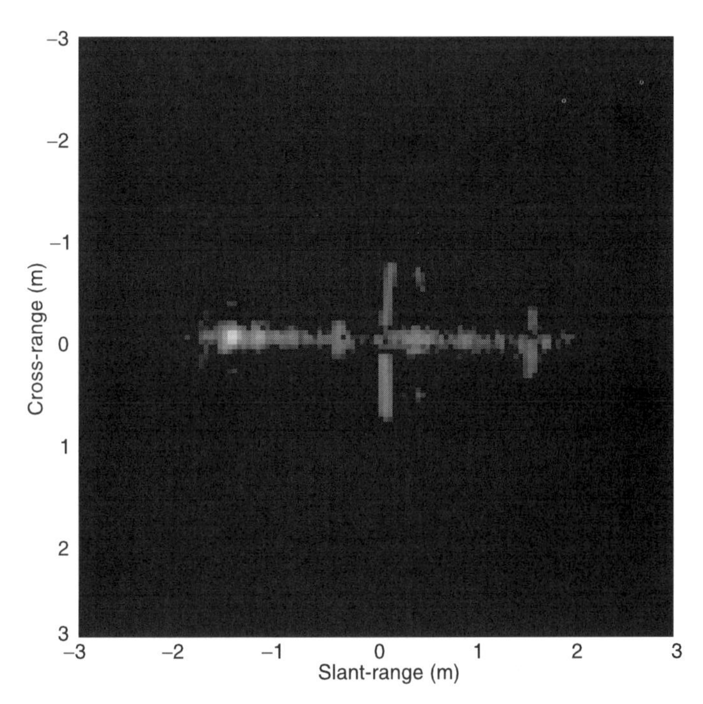


**Figure 13.7** Image of target signature interpolated via cubic splines. (Experimental data courtesy of Naval Air Warfare Center, Point Mugu, California.)

$$\begin{bmatrix} 3h_0 + 2h_1 & h_1 & 0 & \Lambda & 0 \\ h_1 & 2(h_1 + h_2) & h_2 & \Lambda & 0 \\ 0 & h_2 & 2(h_2 + h_3) & h_3 & \mathrm{M} \\ \mathrm{M} & \mathrm{M} & \mathrm{M} & \mathrm{O} & \mathrm{M} \\ 0 & 0 & \Lambda & h_{n-2} & 2h_{n-2} + 3h_1 \end{bmatrix} \begin{bmatrix} S_1 \\ S_2 \\ S_3 \\ \mathrm{M} \\ Sn - 1 \end{bmatrix}$$

$$= 6 \begin{bmatrix} f[x_1, x_2] - f[x_0, x_1] \\ f[x_2, x_3] - f[x_1, x_2] \\ f[x_3, x_4] - f[x_2, x_3] \\ \mathrm{M} \\ f[x_{n-1}, x_n] - f[x_{n-2}, x_{n-1}] \end{bmatrix} \quad (13.5)$$

where

$$f[x_2, x_3] = (y_{i+1} - y_i)/h_i \quad (13.6)$$

Here, $x_i$, $y_i$ are the position and the amplitude of the $i$th sample, and $h_i$ is the width of the $i$th interval, defined as $h_i = (x_{i+1}, x_i)$. The not-a-knot condition makes the first segment equal to the second segment and the last one equal to the one prior to it (i.e., $S_o = S_1$, $S_n = S_{n-1}$), as discussed in [5].

### 13.2.3 Shannon Reconstruction

The Whittaker-Kotel'nikov-Shannon (WKS) sampling theorem [6, 7] states that if a function $f(t)$ is band-limited with a maximum frequency $f_{\max}$, reconstruction can be achieved if the samples are located apart no more than $1/f_{\max}$. The interpolation algorithm based on sinc functions is known as Shannon reconstruction [8] and is defined as

$$f(x) = \sum g(x_n) \operatorname{sinc}\left(\frac{x - x_n}{\Delta}\right) \quad (13.7)$$

where $\operatorname{sinc}(y) = \sin(y)/(y)$. In (3.17), $x_n$ and $g(x_n)$ are the positions and values of the $n$ samples, respectively, and $\Delta$ is the sample spacing. Figure 13.8 shows an image using the Shannon reconstruction method for rotational motion compensation.

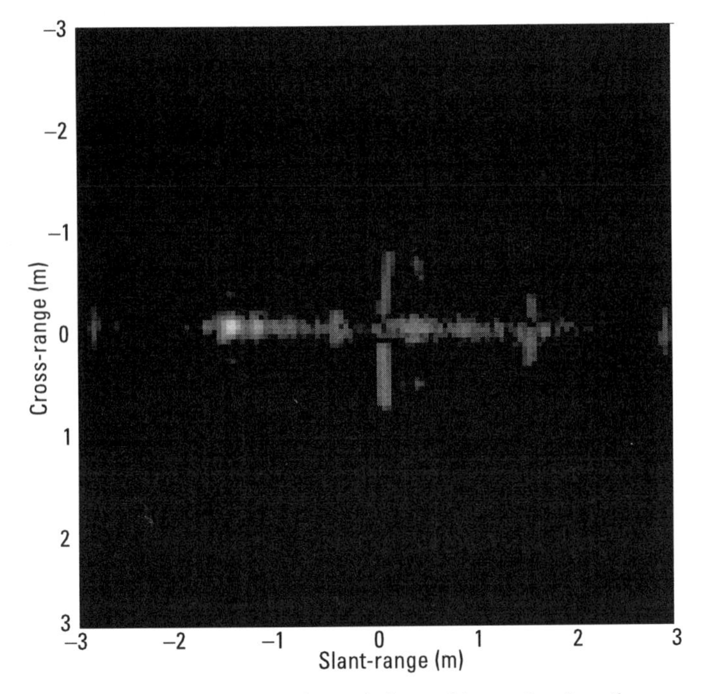


**Figure 13.8** Image obtained from experimental data with rotational motion compensation based on Shannon reconstruction. (Experimental data courtesy of Naval Air Warfare Center, Point Mugu, California.)

### 13.2.4 Sampling Jitter

The Shannon reconstruction algorithm assumes that the samples are evenly spaced with interval $\Delta$. As discussed in Chapter 12, the first step in the interpolation process converts the original polar grid into an equiangular grid. This is done by linearly interpolating between two equiradial points. Note that the interpolated point will no longer be exactly located over equiradial traces. However, for the second interpolation step, the Shannon reconstruction algorithm expects the samples from the first interpolation step to be evenly spaced in the radial direction. The fact that the samples are not evenly spaced can be viewed as a form of sampling jitter. From the standpoint of signal processing, jitter introduces an error that is roughly zero-mean Gaussian [9].

## 13.3 Weighted Integration Techniques

Weighted integration techniques are computationally expensive approaches that consider every sample in the two-dimensional array to interpolate to an unknown point, using kernels such as Bessel and two-dimensional sinc functions.

### 13.3.1 Airy Kernel

The process of sampling in the spatial frequency domain generates an unevenly spaced, band-limited set of data. One of the proposed approaches to solve the problem of image reconstruction from those samples includes Fourier reconstruction methods [10] with explicit polar to rectangular coordinate transformation. Along those lines, Soumekh [8, 11] proposed the use of the following interpolation formula:

$$G(f_x, f_y) = \Delta_\rho \Delta_\theta \sum_{m,n} \frac{G(f_{xmn}, f_{ymn})}{T'(f_{xmn}, f_{ymn})} I(f_x - f_{xmn}, f_y - f_{ymn}) \quad (13.8)$$

where

$$I(f_x, f_y) = X_0^2 J_1 \frac{(\sqrt{f_x'^2 + f_y'^2} X_0)}{\sqrt{f_x^2 + f_y^2} X_0} \quad \text{and} \quad X_0 = \pi/\Delta_\rho \quad (13.9)$$

Equation (13.8) is an Airy function that works as an interpolation kernel. $\Delta_\rho$ and $\Delta_\theta$ correspond to the radial and angular increments. The denominator of the interpolation formula represents the Jacobian of a polar transformation, defined as

$$\left| T'(f_x, f_y) = 1/\sqrt{f_x^2 + f_y^2} \right| \quad (13.10)$$

Equation (13.10) assumes that the desired reconstruction will be bounded within a circular region of radius $X_0$. The reconstructed image is obtained by performing a two-dimensional inverse DFT of an interpolated sample grid with uniform sample spaces $\Delta f_x$ and $\Delta f_y$. Figure 13.9 shows the image compensation obtained by using Airy interpolation for rotational motion compensation.

### 13.3.2 Spatial Domain Reconstruction

According to Soumekh [11], it is possible to interpret (13.8) in the spatial domain as

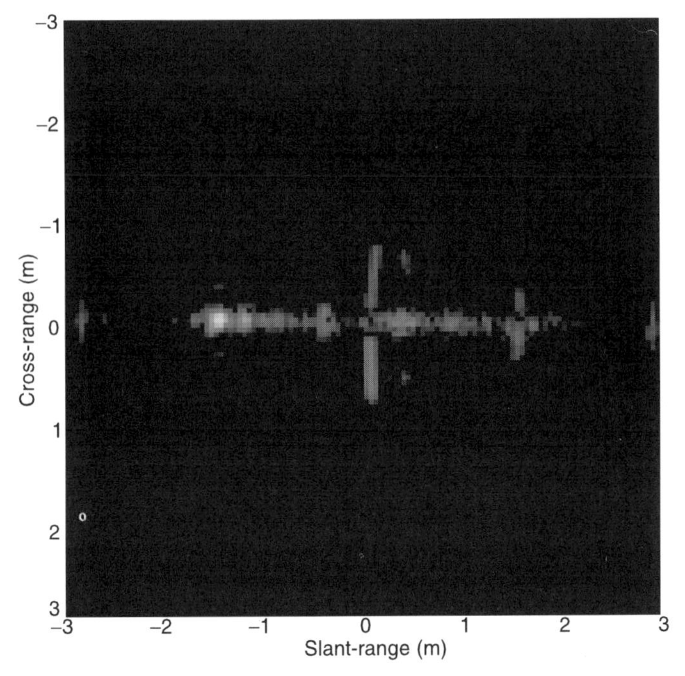


**Figure 13.9** Image of target signature interpolated via Airy function. (Experimental data courtesy of Naval Air Warfare Center, Point Mugu, California.)

$$g(x, y) = \Delta_\rho \Delta_\theta \sum_{m,n} \frac{G(f_{xmn}, f_{ymn})}{|T'(f_{xmn}, f_{ymn})|} \exp(jf_{xmn}x + jf_{ymn}y) \quad (13.11)$$

Using that formula, the image is obtained directly by performing the summation over an evenly spaced grid in the spatial domain. Note that (13, 11) is equivalent to an interpolation operation and a Fourier transformation done simultaneously. The corresponding algorithm uses an exponential kernel and the Jacobian of the polar transformation to interpolate the data.

## 13.4 Comparison of Results Using Various Interpolation Methods

To compare the quality of images reconstructed using the algorithms presented in this section, Matlab® simulations of real data from a U.S. Navy drone model were employed. A total of six algorithms were applied to the data.

### 13.4.1 Simulation Model

A multiple-point target model was used to examine the performance of the interpolation algorithms. The model is based on a drone developed by the Naval Air Warfare Center. The echo return from a target made of $n$ scattering points with a radar cross-section (RCS) of $A_n$ and spatial polar coordinates $(r_n, \phi_n)$ was generated according to the expression

$$G(\rho, \theta) = \sum_n A_n \exp[j4r_n\rho \cos(\theta + \phi_n)/c] \tag{13.12}$$

where $\rho$ and $\theta$ are the frequency domain polar variables. The simulation parameters are in Table 13.1:

### 13.4.2 Image Entropy

In some applications, the amplitude of the scattering points may not be as significant as their position and degree of focus. For a target with multiple scattering points, the spreading of the point response is a significant issue. Therefore, the amount of image spreading introduced by an algorithm is of merit for performance evaluation. In order to determine the amount of spreading created by each interpolation algorithm, the image entropy is calculated in each case [12].

### 13.4.3 Computational Efficiency

Because processing speed is essential for some applications, computational efficiency is an important issue in the selection of a reconstruction algorithm. To

**Table 13.1**
Radar and Simulation Parameters

| Parameter | Value |
|---|---|
| Number of bursts | 128 |
| Pulses per burst | 128 |
| Initial frequency | 8.0125 GHz |
| Bandwidth | 3.175 GHz |
| Aspect change | 18.89 degrees |

evaluate the computational efficiency of an algorithm, the execution time and the number of floating point operations required by Matlab® were determined. A Hewlett-Packard Vectra computer with a 200-MHz Pentium Pro processor was used. Table 13.2 shows the number of flops and the time consumed in the simulations [13].

## 13.5 Rotational Motion Estimation

To generate synthetic sample coordinates of the frequency space signature for a target experiencing continuous rotational motion, a model representative of this motion must first be chosen. The preferred model is quadratic in time given in the form of (12.39). Using this model, the aspect change experienced during the data acquisition time $T$ is $\Delta\theta = \theta(T) - \theta_0$. Prior to image formation, estimates for the angular velocity and acceleration are needed to interpolate the target's frequency response samples from a continuous polar grid to a rectangular grid. It is clear from (12.39) that there exists a multiplicity of values for $\omega$ and $\alpha$ that would satisfy the overall aspect change $\Delta\theta$ for a dwell time $T$. This dilemma requires a solution that can accurately and optimally estimate values for $\omega$ and $\alpha$. The ultimate formation of a focused range-doppler image depends on how close those values are to the true motion parameter values of the target. Estimates deviating from the true motion parameters would yield unfocused imagery, while good estimates would produce focused imagery. Figure 13.10 shows the entropy of the image using various values for rotational motion compensation. The figure clearly shows that the entropy is minimum when the rotating target is compensated with the actual angular velocity.

**Table 13.2**
Computational Efforts for Different Interpolation Methods

| Interpolation method | Flops | Time (seconds) |
|---|---|---|
| Nearest neighbor | 9,207,808 | 35.64 |
| Euclidean | 9,912,320 | 34.50 |
| Linear | 1,668,528 | 75.03 |
| Cubic spline | 5,932,144 | 148.41 |
| Shannon | 63,763,751 | 73.93 |
| Airy | 66.641e9 | 29,338 |

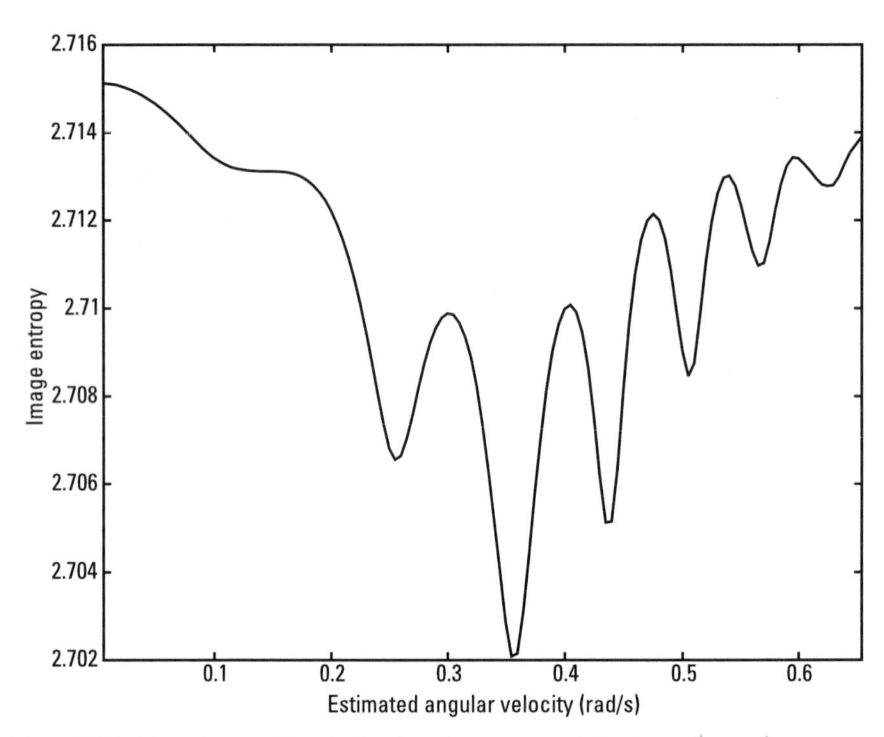


**Figure 13.10** The entropy of the rotational motion compensated using various values.

An effective rotational motion compensation procedure can be summarized as follows. Rotational motion compensation requires the frequency space aperture samples to be polar reformatted using initial motion estimates. The two-dimensional inverse DFT is then computed for the image formation and the image entropy is subsequently measured. In each iteration, new motion parameters are estimated, the data samples are polar reformatted, and the image entropy is calculated. Then the minimum value of the entropy measures and corresponding motion parameter are sought. Using these estimates, the motion compensation step is performed one final time to generate the optimally focused image. The procedure is shown in flow-chart form in Figure 13.11.

## References

[1] Mersereau, R. M., and A. V. Oppenheim, "*Digital Reconstruction of Multidimensional Signals from Their Projections,*" *Proceedings of the IEEE,* Vol 62, No. 10, Oct. 1974, pp. 1319–1338.

[2] Jain, A. K., *Fundamentals of Digital Image Processing,* Prentice Hall, Englewood Cliffs, NJ, 1989.

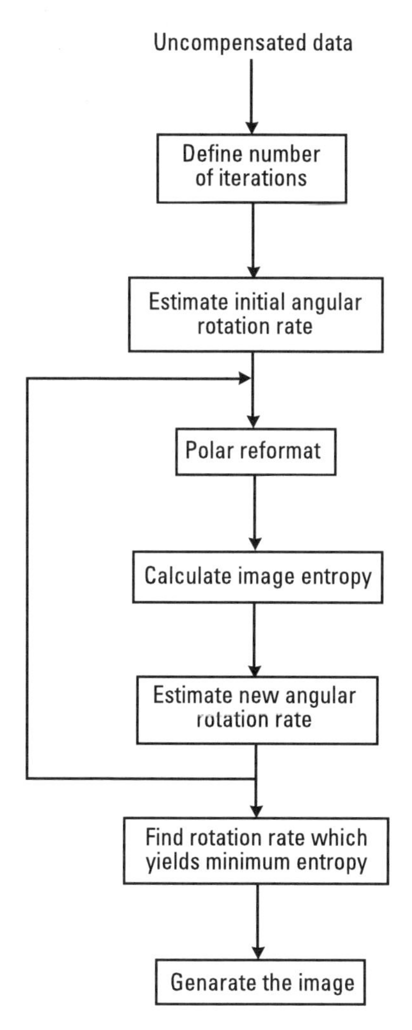


**Figure 13.11** Rotational motion compensation procedure.

[3] Gerald, C. F., and P. O. Wheatley, *Applied Numerical Analysis*, 5th ed., Reading, MA: Addison-Wesley, 1996.

[4] Asaithambi, N. S., *Numerical Analysis, Theory and Practice*, Orlando, FL: Saunders College Publishing, 1995.

[5] De Boor, C., *A Practical Guide to Splines, Applied Mathematical Sciences 27*, New York, NY: Springer-Verlag, 1978.

[6] Clark, J. J., M. R. Paimer, and P. D. Lawrence, "A Transformation Method for the Reconstruction of Functions From Nonuniformly Spaced Samples," *IEEE Transactions on Acoustic, Speech and Signal Processing*, vol. ASSP-33, No. 4, Oct. 1985, pp. 1151–1165.

[7] Jerri, A. J., "The Shannon Sampling Theorem—Its Various Extensions and Applications: A Tutorial Review," *Proceedings of the IEEE*, vol. 65, No. 11, Nov. 1977, pp. 1565–1596.

[8] Soumekh, M., "Band-Limited Interpolation From Unevenly Spaced Sampled Data," *IEEE Transactions on Acoustic, Speech and Signal Processing*, vol. 36, No. 1, Jan. 1988, pp. 110–122.

[9] Scheer, J. A., and J. L. Kurtz, ed., *Coherent Radar Performance Estimation*, Boston, Ma: Artech House, 1993.

[10] Herman, G. T., *Image Reconstruction from Projections: The fundamentals of Computerized Tomography*, New York, NY: Academic Press, 1980.

[11] Soumekh, M., *Fourier Array Imaging*, Englewood Cliffs, NJ: Prentice Hall, 1994.

[12] Wehner, D. R., *High Resolution Radar*, Norwood, MA: Artech House, 1987.

[13] Rojas, R. V., *Comparison of Interpolation Methods for ISAR Imaging*, Master of Science Thesis, University of Texas at El Paso, 1997.

# 14

# Image Enhancement Via Sidelobe Apodization

SAR/ISAR image processing involves a two-dimensional Fourier transform that may produce high intensity sidelobes that obscure low-intensity scatterers in the image. Tapered windows are commonly used, at the cost of decreased resolution to reduce the sidelobe level. Spatially variant sidelobe apodization is a technique that reduces sidelobe levels in a final Fourier image while maintaining the resolution that would be obtained using the rectangular window. This chapter is an introduction to sidelobe apodization and a generalization of the technique based on the use of different parametric windows. When parametric windows are used, low sidelobe levels are obtained at the expense of increasing the complexity of the sidelobe apodization algorithm. Similar resolution and lower sidelobe levels are obtained using a one-dimensional example and a two-dimensional ISAR image.

## 14.1 Introduction to Sidelobe Apodization

Spatially variant apodization (SVA) has been proposed [1–3] for sidelobe level reduction in Fourier imaging. The technique is based on the use of the cosine-on-pedestal weighting function to reduce the sidelobe level from one frequency bin to the next. The weighting function is described as

$$w(n) = 1 + 2\alpha_c \cos(2\pi n/N) \tag{14.1}$$

and its discrete Fourier transform is defined by

$$W(k) = \delta_{k,0} + \alpha_c(\sigma_{k,-1} + \delta_{k,1}) \tag{14.2}$$

for a window length $N$, where

$$\delta_{k,j} = \begin{cases} 1, & k = j \\ 0, & k \neq j \end{cases} \tag{14.3}$$

Equation 14.2 indicates that the implementation of that window reduces to the addition of the amplitude in a selected frequency bin to the intensity of the adjacent pixels, which are weighted by the parameter $\alpha_c$.

It can be seen from (14.1) that the rectangular window ($\alpha_c = 0$) as well as the Hann window ($\alpha_c = 0.5$) are viable options. Thus, the selection of $\alpha_c$ affects the resolution and sidelobe levels. To better understand the SVA approach, a simple form of sidelobe apodization is discussed next.

Applying the rectangular and Hann windows to a time sequence produces two different spectra. Figure 14.1 is an example that uses a multicomponent signal. Components are located at the following normalized frequencies: 0.25, 0.28, and −0.2513. Amplitudes are 1, 0.1, and 0.00001 respectively. Figure 14.1(a) shows a spectrum for which its corresponding data has been weighted using the rectangular window. Figure 14.1(b) shows the case for data weighted using the Hann window. Figure 14.1(c) shows a spectrum generated by comparing the two spectra and selecting the minimum value for each frequency bin. The new spectrum combines the good resolution offered by the rectangular window and the low sidelobe levels obtained by using the Hann window. This approach is known as dual apodization.

## 14.2 Introduction to SVA

Let us define an original complex-valued image as [1]

$$g(m) = I(m) + iQ(m) \tag{14.4}$$

Convolving the image with the frequency domain weighting function defined in (14.2) yields the filtered image

$$g'(m) = \alpha_c(m)g(m-1) + g(m) + \alpha_c(m)g(m+1) \tag{14.5}$$

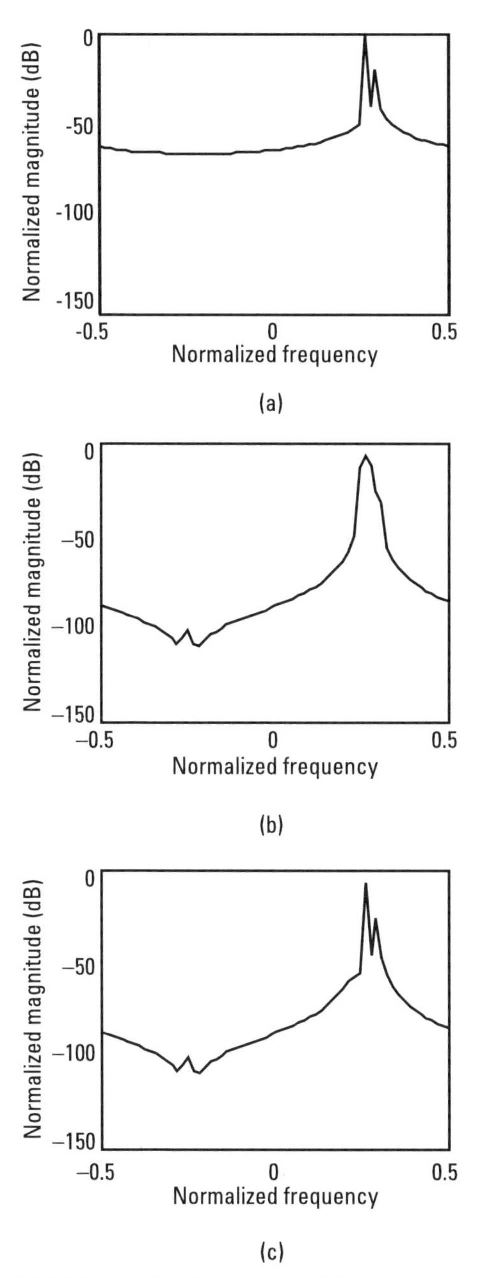


**Figure 14.1** Example of sidelobe apodization using a multi-component signal. The signal consists of three components located at the following normalized frequencies 0.25, 0.28, and −0.2513. Amplitudes are 1, 0.1, and 0.00001 respectively. Figures (a) and (b) show the spectrum obtained when using the rectangular and Hann windows. Dual apodization is presented in (c).

where $\alpha_c$ changes with respect to $m$, so that the best window to be used with the two neighbors of $g(m)$ can be chosen. The parameter $\alpha_c$ can be calculated by minimizing

$$\begin{aligned}|g'(m)|^2 &= \{I(m) + \alpha_c(m)[I(m-1) + I(m+1)]\}^2 \\ &\quad + \{Q(m) + \alpha_c(m)[Q(m-1) + Q(m+1)]\}^2 \qquad (14.6)\end{aligned}$$

subject to the constraint $0 \leq \alpha_c(m) \leq 1/2$ which forces the shape of the window to be within the rectangular and Hann window shapes. Taking the partial derivative of $|g'(m)|^2$ with respect to $\alpha_c$ yields the expression:

$$\begin{aligned}\frac{\partial}{\partial \alpha_c(m)}|g'(m)|^2 &= 2\{I(m) + \alpha_c(m)[I(m-1) + I(m+1)]\} \\ &\quad \times [I(m-1) + I(m+1)] \\ &\quad + 2\{Q(m) + \alpha_c(m) + \alpha_c(m)[Q(m-1) + Q(m+1)]\} \\ &\quad \times [Q(m-1) + Q(m+1)] \qquad (14.7)\end{aligned}$$

Setting the expression equal to zero and solving for the value $\alpha_c$ that minimizes the image intensity around $g(m)$ results in [1]:

$$\begin{aligned}\alpha_c(m) &= \\ &-\frac{\{I(m)[I(m-1) + I(m+1)] + Q(m)[Q(m-1) + Q(m+1)]\}}{[I(m-1) + I(m+1)]^2 + [Q(m-1) + Q(m+1)]^2} \\ &= -\mathrm{Re}\left[\frac{g(m)}{g(m-1) + g(m+1)}\right] = \frac{-|g(m)|}{|g(m-1) + g(m+1)|} \\ &\times \cos\{\arg[g(m)] - \arg[g(m-1) + g(m+1)]\} \qquad (14.8)\end{aligned}$$

where arg[$g$] denotes the phase of $g$.

Taking into consideration the constraint $0 \leq \alpha_c(m) \leq 1/2$, the modified image has the form

$$g'(m) = \begin{cases} g(m), & \alpha_c(m) < 0 \\ g(m) + \alpha_c(m)[g(m-1) + g(m+1)], & 0 \le \alpha_c(m) \le 1/2 \\ g(m) + (1/2)[g(m-1) + g(m+1)], & \alpha_c(m) > 1/2 \end{cases} \tag{14.9}$$

For closely spaced signal components located around pixel $m$, the values of $\alpha_c$ are close to or equal to zero. Thus, the operation is equivalent to applying the rectangular window. Otherwise, the value for $\alpha_c$ is calculated so that the lowest possible sidelobe is obtained by considering a large number of cosine-on-pedestal window shapes. When $\alpha_c > 0.5$, the Hann window is applied as stated in (14.9).

Figure 14.2 shows the results using the SVA approach based on the cosine-on-pedestal window. Note how the SVA method yields the lowest sidelobe levels when different windows obtained by varying the value of $\alpha_c$ in (14.9) are used.

Using the real $I(m)$ and imaginary $Q(m)$ channels of $g(m)$ separately yields a simpler implementation of the SVA method. When a change of sign occurs in $g'(m)$, a real valued weighting function given by

$$\alpha_c(m) = \frac{-g(m)}{g(m-1) + g(m+1)} \tag{14.10}$$

for which $g'(m) = 0$. Thus, a new condition in (14.9) is given by

$$g'(m) = 0 \quad \text{for} \quad 0 \le \alpha_c(m) \le 1/2 \tag{14.11}$$

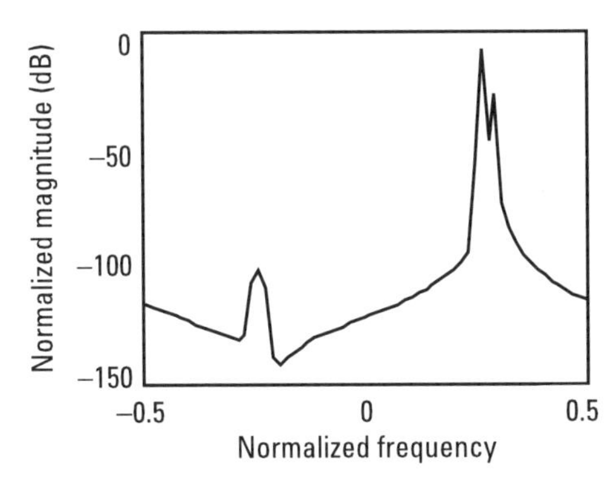


**Figure 14.2** Results using the SVA technique with the same example as in Figure 14.1. Lower sidelobe levels and better resolution are obtained.

An example of sidelobe apodization processing using the *I* and *Q* channels separately is shown in Figure 14.3. In this figure, the same signal example is used. Note how the condition in (14.11) reduces sidelobes completely in most of the areas where sidelobes are predominant.

A complete description of the method for the two-dimensional case can be found in Stankwitz et al [1].

## 14.3 Parametric Windows

Five windows have similar parameter dependence are:

- The Tukey window, defined as [4]

$$w(n) = \begin{cases} 1, & 0 \leq |n| \leq \alpha_T \dfrac{N}{2} \\ 0.5\left[1 + \cos\left(\pi \dfrac{n - \alpha_T(N/2)}{2(1-\alpha_T)(N/2)}\right)\right], & \alpha_T \dfrac{N}{2} < |n| \leq \dfrac{N}{2} \end{cases} \quad (14.12)$$

- The Kaiser window [5], defined as

$$w(n) = \frac{I_0\left(\pi \alpha_K \sqrt{1 - \left(\dfrac{n}{N/2}\right)^2}\right)}{I_0(\pi \alpha_K)}, \quad 0 \leq |n| \leq N/2 \quad (14.13)$$

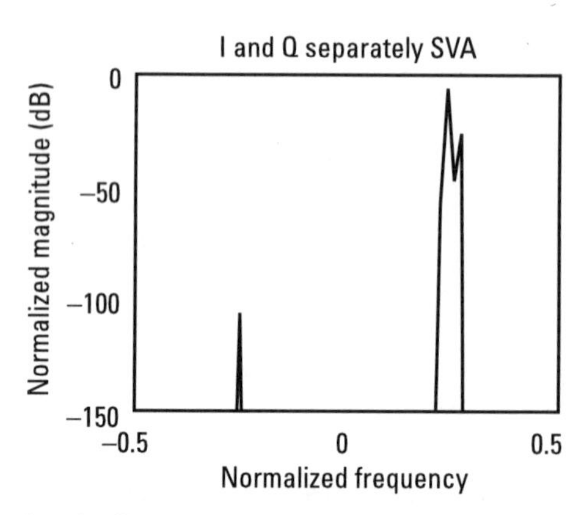


**Figure 14.3** Results using the SVA technique with the *I* and *Q* channels separately.

where $I_0(x)$ denotes the modified zeroth order Bessel function

$$I_0(x) = \sum_{k=0}^{\infty} \left[ \frac{\left(\frac{x}{2}\right)^k}{k!} \right]^2 \tag{14.14}$$

- The Poisson window, defined as [4]

$$w(n) = \exp\left(-\alpha_P \frac{|n|}{N/2}\right), \qquad 0 \le |n| \le N/2 \tag{14.15}$$

- The Gaussian window, defined as [4]

$$w(n) = \exp\left[-\frac{1}{2}\left(\alpha_G \frac{n}{N/2}\right)^2\right], \qquad 0 \le |n| \le N/2 \tag{14.16}$$

- The Cauchy family of windows, defined as [4]

$$w(n) = \frac{1}{1 + \left[\alpha_C \frac{n}{N/2}\right]^2}, \qquad 0 \le |n| \le N/2 \tag{14.17}$$

Those windows assume the shape of the rectangular window as the $\alpha$'s tend to zero, as shown in Figure 14.4, where multiple windows are formed using different values for $\alpha$. The use of the rectangular window yields optimum resolution on the resulting spectrum when sidelobe apodization is implemented using one of the windows. The best $\alpha$ parameters for sidelobe control are found by minimizing the spectral response on a pixel-by-pixel basis.

Using those windows and applying the same concept as dual apodization, the results for the same example used in this chapter are shown in Figure 14.5.

## 14.4 Sidelobe Apodization Using the Kaiser Window

As noted in [2], SVA is a version of the minimum variance spectral estimator with a restricted set of weighting functions using a simple estimate of the covariance function.

Other weighting functions (Poisson, Gaussian, Tukey, Kaiser, and Cauchy) can be used as an estimate of the covariance function for sidelobe apodization purposes [6]. Among those weighting functions, the use of the Kaiser window

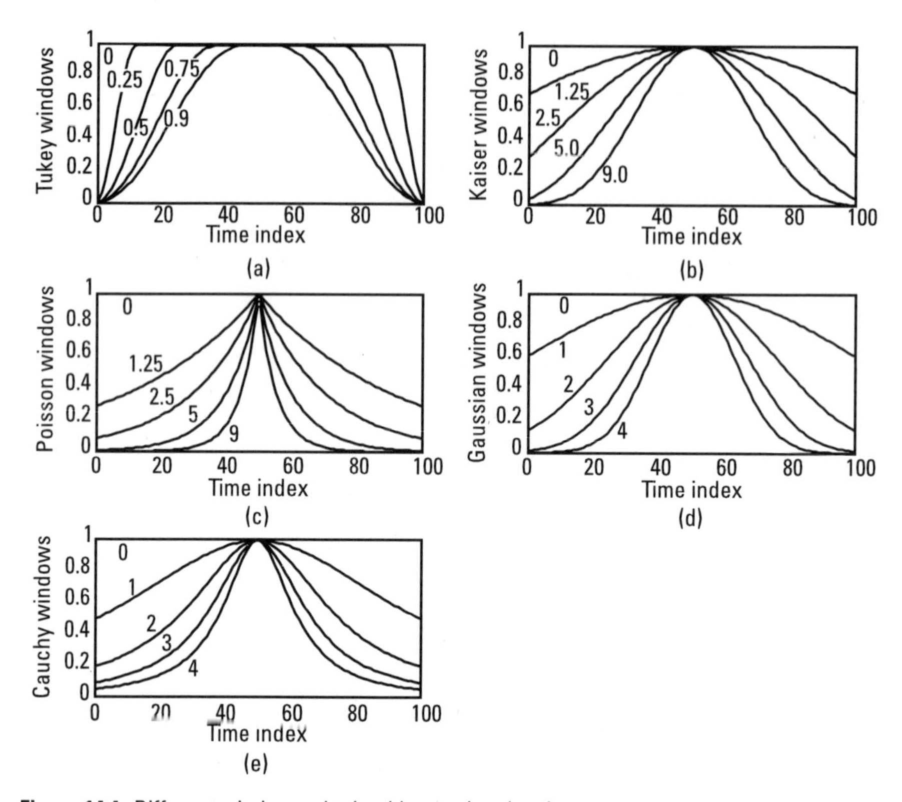


**Figure 14.4** Different windows obtained by varying the shape parameter.

has shown excellent results when compared with the SVA method. The comparison is between the SVA using the $I$ and $Q$ channels jointly as stated in (14.9) and the absolute value of the spectra when multiple Kaiser windows are used. Thus, an approach similar to that presented in [1] to achieve sidelobe apodization using the Kaiser window is formulated.

The DFT of the Kaiser window can be expressed as [4]

$$W(k) = \frac{N}{I_0(\alpha_K \pi)} \frac{\sinh\left[\sqrt{\alpha_K^2 \pi^2 - (Nk/2)^2}\right]}{\sqrt{\alpha_K^2 \pi^2 - (Nk/2)^2}}$$

$$\text{for} \quad k = -\pi, \ldots, -2\pi/N, 0, 2\pi/N, \ldots, \pi \quad (14.18)$$

Thus, the use of $W(k)$ in the image domain using a convolution is not as simple as in the case of using the Kronecker delta functions in the SVA method.

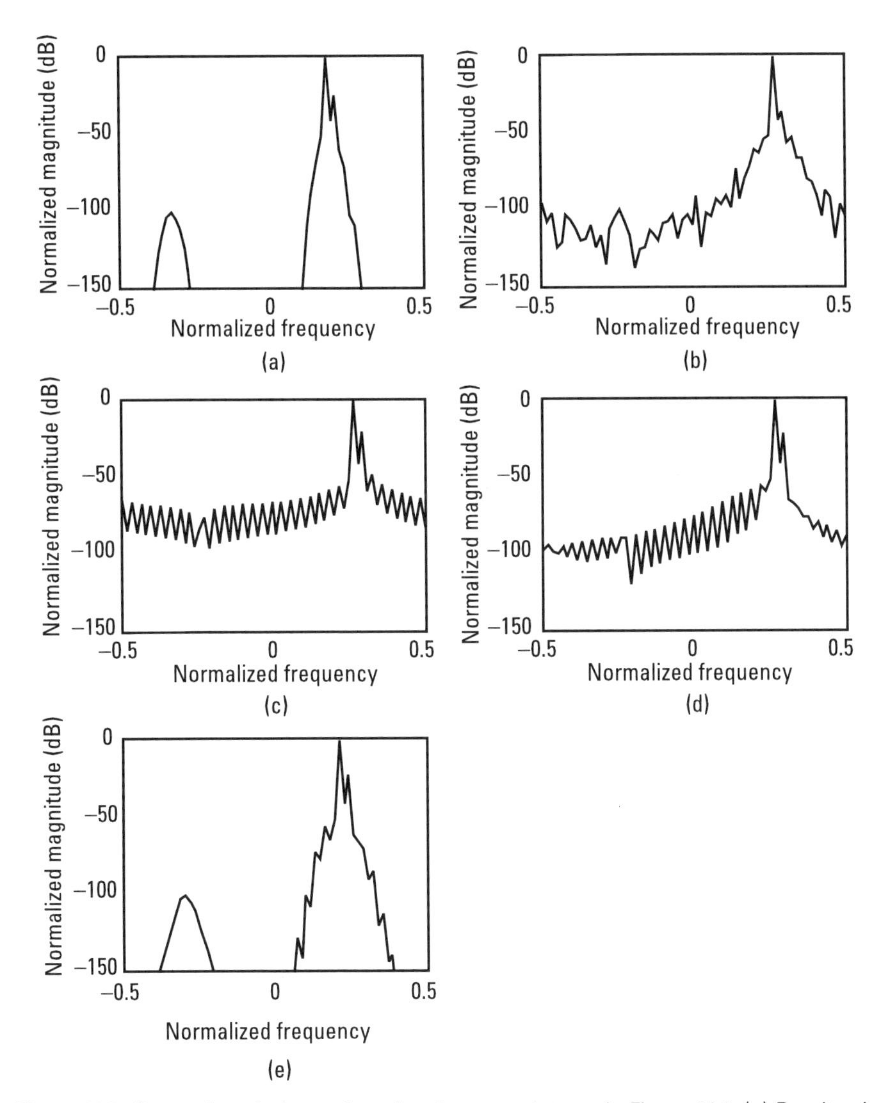


**Figure 14.5** Spectral analysis results using the same data as in Figure 14.1. (a) Results obtained using the Kaiser window approach; (b) results of the Tukey window approach; (c) results of the Poisson window technique; (d) results of the Cauchy window; (e) results of the Gaussian window approach.

Kaiser found the following empirical relations between the sidelobe attenuation levels ($ATT$) and the value of $\alpha_K$ [5]:

$$\alpha_K = \begin{cases} 0.1102(ATT - 8.7). \quad ATT > 50 \\ 0.5842(ATT - 21)^{0.4} + 0.07886(ATT - 21), \quad 21 \leq ATT \leq 50 \\ 0.0, \quad ATT < 21 \end{cases} \tag{4.19}$$

From (14.19), it can be seen that for $\alpha_K > 30$, an attenuation greater than 300 dB is possible. With this in mind, and setting this attenuation as a boundary, only few samples around $k = 0$ in (14.18) contribute considerably to the output of the convolution implementation of the weighting function, as shown in Figure 14.6.

Thus, we can modify the image and implement the convolution as

$$g'(m) = \sum_{i=-n}^{n} W(2\pi i/N)g(m - i) \tag{14.20}$$

with just a few samples of (14.18). For $n = 1$, the same number of Kronecker delta functions will be used as in the SVA approach. When we use $n = 1, 2, 3, 4$, better results are obtained.

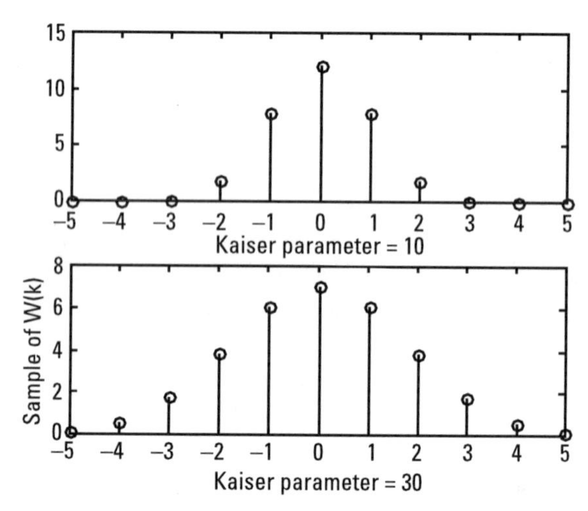


**Figure 14.6** Samples of $W(k)$ in (14.4), for $\alpha_K = 10$, and $\alpha_K = 30$.

Table 14.1 shows the ratio

$$100 \times \sum_{i=-n}^{n} |W(2\pi i/N)|^2 / \sum_{i=-N/2}^{N/2} |W(2\pi i/N)|^2 \qquad (14.21)$$

for $N = 32$, using different values of $n$ and $\alpha_K$. From Table 14.1, similar results can be expected when most of the samples given in (14.18) are discarded as compared to applying Kaiser multiapodization, which would use all the samples.

One point to note is that the best $\alpha_K$ must be found for every pixel. Thus, solving for $|g'(m)|^2 = 0$ using (14.20) requires the use of a numerical analysis technique to find the value of $\alpha_K$ that guarantees a minimum sidelobe level. Even though only a few samples are used in (14.18), the complexity of the equation compared with the Kronecker delta functions in the SVA approach makes the Kaiser window approach slow.

Figure 14.7 shows the spectrum of the signal weighted with the rectangular window (solid line). Next, the figure shows the case for data weighted using the Hann window. Other results correspond to the use of the SVA and Kaiser windows when using $n = 1$, and $n = 2$. As can be seen, the Kaiser window offers the best sidelobe reduction while maintaining thc resolution offered by the rectangular window (no apodization).

Using the SVA and Kaiser approaches processing the *I* and *Q* channels separately, yield the same results shown in Figure 14.3. The use of the Kaiser window approach for this case is not desirable since the number of computations needed are far greater than those required by the SVA technique and the results are similar.

**Table 14.1**

| $\alpha_K$ | $n = 1$ | $n = 2$ | $n = 3$ | $n = 4$ |
|---|---|---|---|---|
| **1** | 99.9654 | 99.9921 | 99.9972 | 99.9988 |
| **10** | 97.3169 | 99.9955 | 99.9999 | 99.9999 |
| **20** | 86.3242 | 98.8102 | 99.9689 | 99.9999 |
| **30** | 77.0097 | 95.5826 | 99.5531 | 99.9787 |

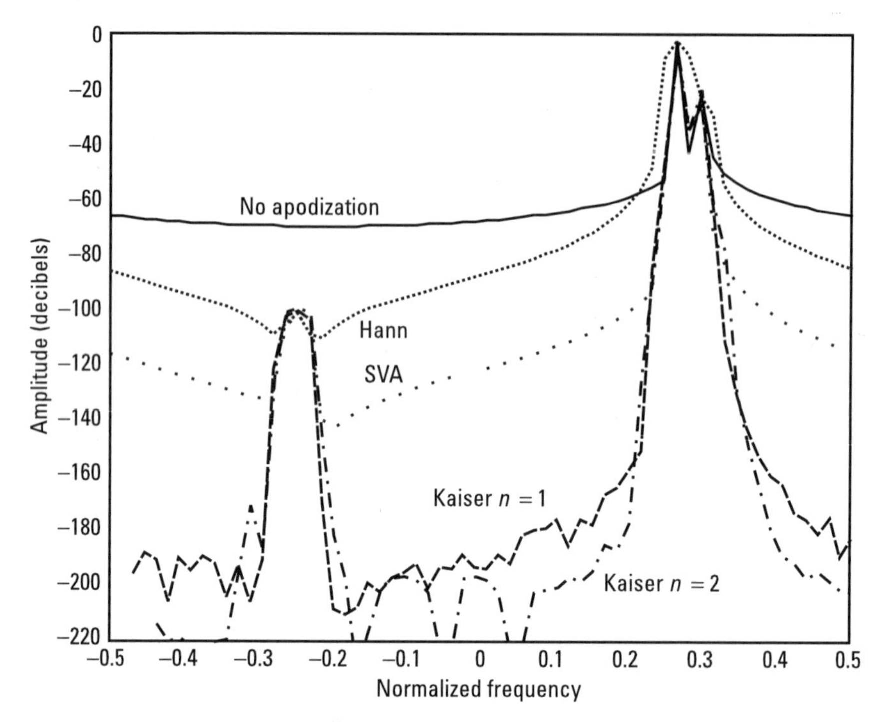


**Figure 14.7** Example using a one-dimensional signal.

In window design, two main figures of merit are the highest sidelobe level (HSL) and the sidelobe fall-off (SFO) [4]. For the Tukey window, those two values are −19 dB and −18 dB/octave ($\alpha_T = 0.75$). For the Kaiser window, those vlues are −82 dB and −6 dB/octave ($\alpha_K = 3.5$) respectively. It can be seen that while the Kaiser window has higher HSL values than the Tukey window, the opposite occurs for the SFO values. From the examples presented in this chapter, it seems that the HSL value is important in order to determine an effective window.

Figure 14.8 shows the spectrum of a noisy signal processed via sidelobe apodization. The test signal is the same as the one used in Figure 14.7 except that zero mean Gaussian noise has been added at the same level of the component located at normalized frequency −0.2513. As expected, the methods reduce sidelobes but not the noise level. Thus, the −100 dB amplitude component can not be discerned.

As indicated in [1], SVA reduces the spectrum spreading presented when the component phase is varying with time. Figure 14.9 shows an example using a simulated point scatterer experiencing rotational motion. The rotation causes blurring on a final image as indicated in Chapter 12. As can be seen from

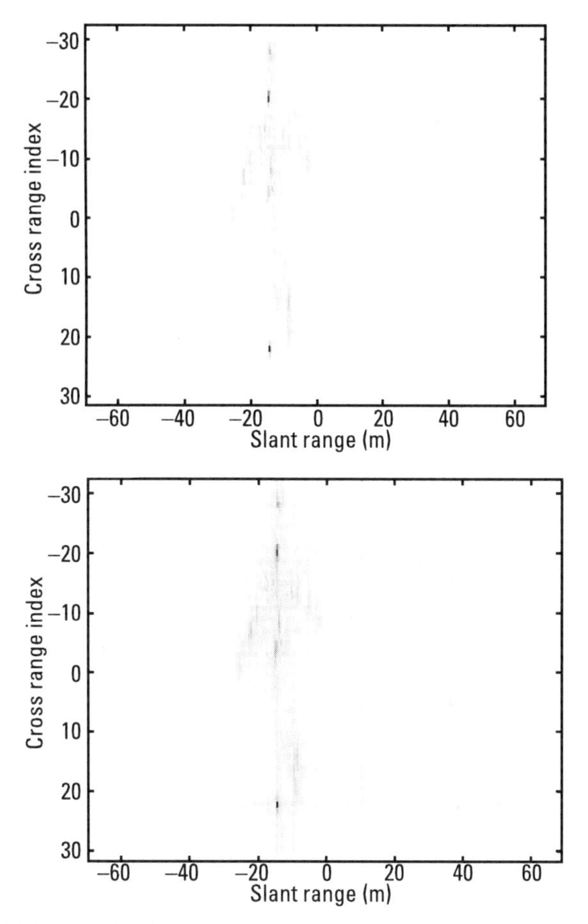


**Figure 14.8** Sidelobe apodization example using a signal with zero mean additive Gaussian noise.

Figure 14.9, the Kaiser window can also has a potential use for motion compensation purposes.

The SVA and Kaiser approaches can be extended to two dimensions by following standard column-row decomposition. Figure 14.10 shows the results when the approach is applied to SAR or ISAR data. The dynamic range between point scatterers in SAR/ISAR is a good example of how sidelobes yield undesired effects on target images. Note that the target's point scatterers are clearly discernible when the Kaiser window is used. That is the result of improvement in the suppression of sidelobes.

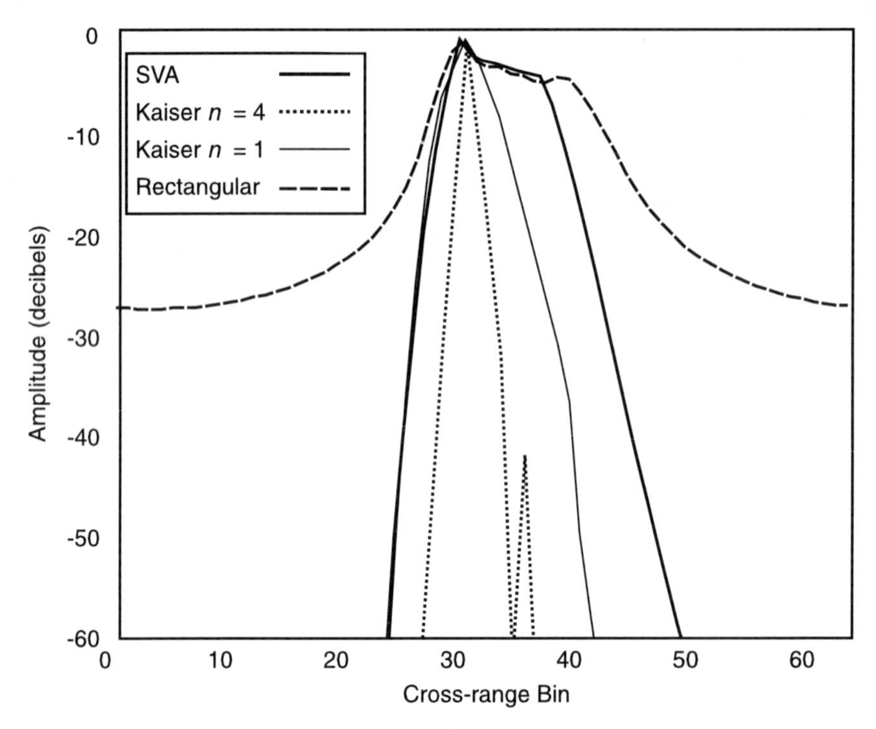


**Figure 14.9** Example using a rotating point scatterer.

## 14.5 Summary

This chapter discussed the use of different parametric windows for sidelobe apodization purposes. Sidelobe apodization using the Kaiser window yields images with suppressed sidelobes when applied to one-dimensional and two-dimensional data. The necessary condition for applying a window other than the Kaiser window is that within the multiple versions of the window proposed the rectangular window be included so that maximum resolution is achieved. In fact, results between different windows can be combined. Although better sidelobe reduction may be achieved, the application of a new window in the Fourier domain may not be as simple as in the case of the cosine-on-pedestal window. Thus, the computation of optimal window parameters may require a considerable computational effort.

## References

[1] Stankwitz, H. C., R. J. Dallaire, and J. R. Fienup, "Non-Linear Apodization for Sidelobe Control in SAR Imagery," *IEEE Trans. on Aerospace and Electronics Systems*, Vol. 31, No. 1, Jan. 1995, pp. 267–279.

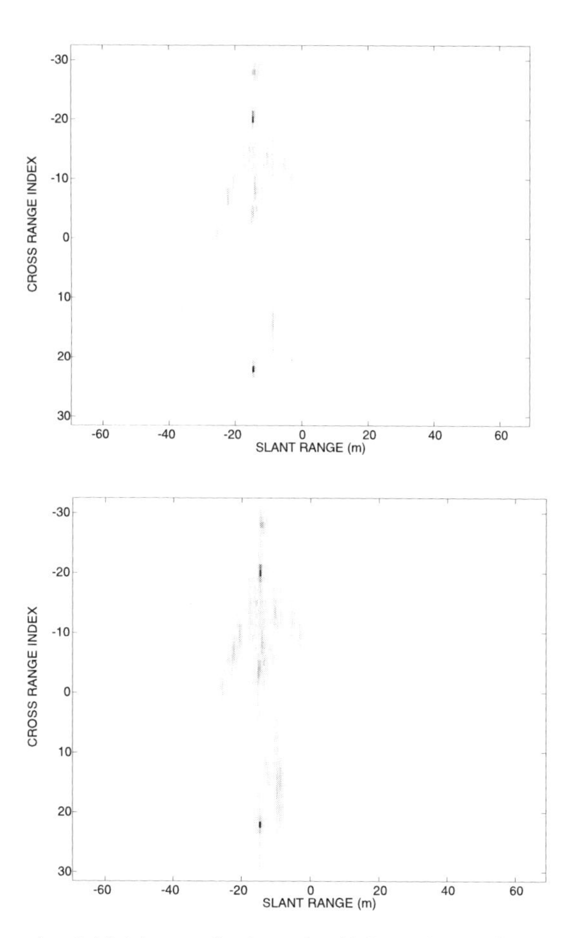


**Figure 14.10** Example of sidelobe apodization using high resolution ISAR data. (a) Image obtained using the Kaiser window approach; (b) image obtained using the SVA technique.


[2] Lee, J. A. C., D. C. Munson, Jr., "Effectiveness of Spatially-Variant Apodization," *Proc. IEEE Internat. Conf. on Image Processing*, Washington, DC, Oct. 1995, pp. 147–150.

[3] Degraaf, S. R., "SAR Imaging via Modern 2D Spectral Estimation Methods," *IEEE Trans. on Image Processing*, Vol. 7, No. 5, May 1998.

[4] Harris, F. J., "On the Use of Windows for Harmonic Analysis With the Discrete Fourier Transform," *IEEE Proc.*, Vol. 66, No. 1, Jan. 1978.

[5] Kaiser, J. F., "Nonrecursive Digital Filter Design Using the Io-Sinh Window Function," *Proc. 1974 IEEE Internat. Symp. on Circuits and Systems*, Apr. 1974, pp. 20–23.

[6] Thomas, G., J. Sok-Son, B. C. Flores, "Sidelobe Apodization Using Parametric Windows," *SPIE Proc. Algorithms for Synthetic Aperture Radar Imagery VI*, Orlando, FL, Vol. 3721, Apr. 1999.

# Index

## Recent Titles in the Artech House Radar Library

David K. Barton, Series Editor

*Advanced Techniques for Digital Receivers,* Phillip E. Pace

*Airborne Pulsed Doppler Radar, Second Edition*, Guy V. Morris and Linda Harkness, editors

*Bayesian Multiple Target Tracking*, Lawerence D. Stone, Carl A. Barlow, and Thomas L. Corwin

*CRISP: Complex Radar Image and Signal Processing, Software and User's Manual,* August W. Rihaczek, et al.

*Design and Analysis of Modern Tracking Systems,* Samuel Blackman and Robert Popoli

*Digital Techniques for Wideband Receivers,* James Tsui

*Electronic Intelligence: The Analysis of Radar Signals, Second Edition,* Richard G. Wiley

*Electronic Warfare in the Information Age*, D. Curtis Schleher

*EW 101: A First Course in Electronic Warfare,* David Adamy

High-Resolution Radar, Second Edition, Donald R. Wehner

*Introduction to Electronic Warfare,* D. Curtis Schleher

*Microwave Radar: Imaging and Advanced Concepts,* Roger J. Sullivan

*Millimeter-Wave and Infared Multisensor Design and Signal Processing,* Lawrence A. Klein

*Modern Radar System Analysis,* David K. Barton

*Multitarget-Multisensor Tracking: Applications and Advances Volume III*, Yaakov Bar-Shalom and William Dale Blair, editors

*Principles of High-Resolution Radar,* August W. Rihaczek

*Radar Cross Section, Second Edition,* Eugene F. Knott, et al.

*Radar Evaluation Handbook*, David K. Barton, et al.

*Radar Meteorology,* Henri Sauvageot

*Radar Signal Processing and Adaptive Systems,* Ramon Nitzberg

*Radar System Performance Modeling*, G. Richard Curry

*Radar Technology Encyclopedia*, David K. Barton and Sergey A. Leonov, editors

*Range-Doppler Radar Imaging and Motion Compensation,* Jae Sok Son, et al.

*Theory and Practice of Radar Target Identification,* August W. Rihaczek and Stephen J. Hershkowitz

For further information on these and other Artech House titles, including previously considered out-of-print books now available through our In-Print-Forever® (IPF®) program, contact:

Artech House
685 Canton Street
Norwood, MA 02062
Phone: 781-769-9750
Fax: 781-769-6334
e-mail: artech@artechhouse.com

Artech House
46 Gillingham Street
London SW1V 1AH UK
Phone: +44 (0)20 7596-8750
Fax: +44 (0)20 7630-0166
e-mail: artech-uk@artechhouse.com

Find us on the World Wide Web at:
www.artechhouse.com